Network Behavior Analysis

Kuai Xu

Network Behavior Analysis

Measurement, Models, and Applications

Kuai Xu
School of Mathematical
and Natural Sciences
Arizona State University
Glendale, AZ, USA

ISBN 978-981-16-8327-5 ISBN 978-981-16-8325-1 (eBook)
https://doi.org/10.1007/978-981-16-8325-1

This Springer imprint is published by the registered company Springer Nature Singapore Pte Ltd.
The registered company address is: 152 Beach Road, #21-01/04 Gateway East, Singapore 189721, Singapore

Preface

As the Internet continues to grow in size and complexity, the challenge of effectively provisioning, managing, and securing it has become inextricably linked to a deep understanding of network behaviors of networked systems and Internet applications. While there exists an extensive body of research publications on traffic classifications, Internet measurement, network security, and digital forensics, there are few books dedicated to network behavior analysis. This book provides a comprehensive overview of network behavior analysis that focuses on the study of network traffic data to provide critical insights into the behavioral patterns of networked systems such as servers, desktops, smartphones, and the Internet of Thing (IoT) devices and Internet applications such as web browsing, electronic mails, file transfers, online gaming, video streaming, and social networking. The objective of this book is to fill the book publication gap in network behavior analysis which has recently become an increasingly important component of comprehensive network monitoring and security solutions for backbone networks, enterprise networks, data center networks, home networks, and emerging networks such as 5G networks, vehicle networks, and IoT networks.

Network behavior analysis is an end-to-end process of collecting, extracting, analyzing, modeling, and interpreting network behavior of end systems and network application from Internet traffic data such as TCP/IP data packets and network flows. This book presents the fundamental principles and best practices for network behavior analysis. Relying on data mining, machine learning, information theory, probabilistic graphical and structural modeling, this book explains what, who, where, when, and why of communication patterns and network behavior of networked systems and Internet applications. The book also discusses the benefits of network behavior analysis for the applications of cybersecurity monitoring, Internet traffic profiling, anomaly traffic detection, and emerging application detection.

This book is of particular interest to researchers and practitioners in the fields of Internet measurement, traffic analysis, cybersecurity since this book brings a spectrum of innovative techniques to develop behavior models, structural models, graphic models of Internet traffic and presents how to leverage these results from these models for a broad range of real-world applications in network management, security operations, and cyber intelligent analysis.

The major benefits of reading this book include (1) learning the principles and practices of measuring, modeling, and analyzing network behavior from massive Internet traffic data; (2) making sense of network behavior for a spectrum of applications ranging from cybersecurity, network monitoring and emerging application detection; and (3) understanding how to explore network behavior analysis to complement traditional perimeter-based firewall and intrusion detection systems to detect unusual traffic patterns or zero-day security threats via data mining and machine learning techniques. The prerequisite for reading this book is a basic understanding on TCP/IP protocols, data packets, network flows, and Internet applications.

Phoenix, AZ, USA
October 2021

Kuai Xu

Acknowledgements

I would like to acknowledge the contributions to the research publications, which build the theoretical and system foundations of this book, from my collaborators and co-authors: Supratik Bhattacharyya, Jianhua Gao, Lin Gu, Yaohui Jin, Yinxin Wan, Feng Wang, Guoliang Xue, and Zhi-Li Zhang.

I would also like to thank Sprint, Center for Applied Internet Data Analysis (CAIDA) based at the University of California's San Diego Supercomputer Center, and University of Oregon Route Views Project for making Internet traffic datasets and routing table datasets available to the research projects in the book.

I am grateful for the constructive feedbacks and insightful comments from the reviewers, which have greatly improved the technical presentation of the book. I also would like to thank Springer editorial staff for their support and encouragement from the beginning of the book project to the final publication.

The research projects in this book are partially supported by three research grants (CNS-1218212, DMS-1737861, and CNS-1816995) from US National Science Foundation (NSF).

Contents

Chapter 1
Introduction

Abstract As the Internet continues to grow in devices, applications, services, users, and traffic, network behavior analysis has increasingly become a crucial area in research for understanding what is happening on the Internet. This chapter first introduces the definition and concept of network behavior and discusses the importance and urgent need of network behavior analysis. Subsequently, this chapter describes the common methods, infrastructure, and frameworks for collecting, monitoring, modeling and analyzing network behavior. Next, this chapter discusses the broad benefits and applications of exploring network behavior in behavioral profiling, anomaly detection, traffic engineering, and security monitoring. Finally, this chapter concludes with an overview of the topics covered in this book and the overall organization of the book chapters.

1.1 What is Network Behavior Analysis

As the Internet continues to grow in size and complexity, the challenge of effectively provisioning, managing, and securing it has become inextricably linked to a deep understanding of Internet traffic and network behavior [1–5]. The imperative and urgency of understanding network behavior and traffic patterns gives rise to the field of network behavior analysis (NBA), which focuses on the study of network traffic data for providing critical insights into behavioral patterns of end systems and network applications.

Throughout this book, the terms, `networked systems`, `end hosts`, and `end systems`, will be used interchangeably for representing *all* Internet-connected devices including desktops, laptops, tablets, smartphones, servers as well as Internet of Thing (IoT) devices. We will also use the terms, `Internet applications`, `network applications`, and `application services` interchangeably for denoting the broad range of applications and services running on the Internet infrastructure, such as e-mail, web, video and music streaming, gaming, online social media, and smartphone apps.

K. Xu, *Network Behavior Analysis*,
https://doi.org/10.1007/978-981-16-8325-1_1

Network behavior analysis is an end-to-end process of collecting, extracting, analyzing, modeling, and interpreting network behavior of end systems and network application from massive amount of Internet traffic data such as TCP/IP data packets and network flows. Network behavior analysis is an interdisciplinary research field that involves a variety of disciplines such as computer science, machine learning, data mining, artificial intelligence, visualizations, statistical analysis, network science, and information theory.

Network behavior analysis includes a suite of models, algorithms, and tools for characterizing analyzing behavioral patterns of networked systems and Internet applications [6, 7]. Network behavior analysis shares a number of methods, techniques, and tools with network forensics which typically focuses on reactively investigating cyber attacks, cyber crimes, threats, and vulnerabilities with network traffic traces. However, network behavior analysis has a much broader range of applications such as network engineering, capacity planning, traffic optimizations, behavior profiling, and security monitoring. In addition, network behavior analysis could be employed proactively or reactively (or both) for these applications.

1.2 Network Behavior Measurement and Modeling

The last two decades have witnessed significant progress in instrumenting data collection systems including Internet measurement and monitoring platforms [8–14] for high-speed networks at the core and edges of the Internet. The measurement and monitoring platforms for network behavior analysis [9, 11, 12] are among these instruments with specific objectives of collecting Internet traffic data for analyzing and modeling behavioral patterns of networked systems and Internet applications.

In light of wide spread cyber attacks [15–20] and the frequent emergence of disruptive applications [21–24] that often rapidly alter the dynamics of network traffic, and sometimes bring down valuable Internet services, network behavior modeling has a primary objective of unveiling the underlying structures and communication patterns from Internet traffic data for use in network operations and security management.

Due to vast quantities of traffic data, diverse networked systems, and Internet applications, characterizing and modeling network behavior from massive network traffic data is a daunting task for researchers and practitioners in both academia and industry. The recent research and development on network behavior analysis [9, 25–32] have introduced three different and complementary models characterizing and interpreting communications patterns of networked systems and Internet applications: behavioral model [9, 33], structural model [34–36], and graphical model [27, 37–39].

The behavioral, structural and graphical models of network traffic collectively and complementarily describe and capture *who*, *what*, *when*, *where*, and *why* of data communications on the Internet. The behavioral model builds behavioral patterns of networked systems and Internet applications via summarizing traffic features of data communications, while the structural model of network traffic captures and characterizes the interactions between various features or dimensions in traffic data as well as identifies the dominant feature values, if exists, for describing the communication structure and activity for the behavioral entities. Rather than focusing on individual network systems and Internet applications, the graphical model of network traffic explores bipartite graphs and one-mode projections for measuring and quantifying behavior similarity of network systems and Internet applications, and discovers the inherent groups or clusters of networked systems and Internet applications exhibiting similar behavioral patterns in data communications.

1.3 Benefits of Network Behavior Analysis

The insights of behavioral patterns resulted from network behavior analysis are of practical value to a wide spectrum of applications in network operations and cybersecurity such as traffic characterization and classification [40–42], security monitoring [43], and network forensics [44, 45].

The network behavior analysis provides critical insights into traffic characterization and classifications for network applications [46], end systems [47, 48], and Internet users [49–51]. The rich set of multidimensional and multi-layer traffic features from network behavior analysis not only characterizes traffic patterns of the Internet "objects", i.e., Internet applications, network systems, and end users but also enables accurate classifications and detection on unknown or anomalous network traffic [52].

The in-depth understanding of normal and abnormal behavioral patterns of Internet applications and network systems supports network operators and security analysts to design and deploy effective security monitoring platforms for continuously monitoring, detecting, alerting, filtering, and mitigating suspicious behaviors and anomalous activities. In general, establishing the baseline behavioral patterns of Internet applications and network systems via longitude network measurement studies leads to effective detection of intrusion activities or anomalous behavior [9, 53, 54].

Network forensics, a subfield of digital forensics, focuses on collecting, monitoring, and analyzing network traffic data for investigation and analysis of cyber attacks and cyber crimes, intrusion detection and prevention, and evidence gathering of data communication over the Internet. The network forensic investigators can leverage network behavior analysis for understanding the behavioral patterns of suspicious attacker systems and the victim targets and for discovering behavioral dynamics from the system data logs and network traffic traces before, during, and after cyber attacks or cyber crimes. Network behavior analysis shares many similarities in network traffic collection and analysis with network forensics analysis. However, the primary objective of network behavior analysis is to uncover behavioral patterns of Internet applications, network systems, and end users, which provide critical insights into network forensics, traffic analysis, and security monitoring. In other words, network forensics is one of the practical applications of network behavior analysis.

1.4 Book Overview and Organization

This book provides a comprehensive overview of network behavior analysis that mines Internet traffic data for extracting, modeling, and making sense of behavioral patterns of Internet "objects", such as smartphones, IoT, servers, web applications, and Internet gaming. The objective of this book is to fill the book publication gap in network behavior analysis, which has recently become an increasingly important component of comprehensive network measurement and monitoring solutions for data center networks, backbone networks, enterprise networks, and home networks.

This book presents the fundamental principles and best practices for measuring, extracting, modeling, and analyzing network behavior from Internet traffic data for network systems and Internet applications. Relying on data mining, machine learning, information theory, probabilistic graphical and structural modeling, this book explains what, who, where, when, and why of communication patterns and network behavior of network systems and Internet applications. The book also discusses the benefits of network behavior analysis for the applications of cybersecurity monitoring, Internet traffic profiling, anomaly traffic detection, and emerging application detection.

The remainder of this book is organized as follows. Chapter 2 presents the background of network behavior analysis. Chapter 3 introduces behavior models of network traffic. Chapter 4 describes structural models of network traffic. Chapter 5 discusses graphical models of Internet traffic. Chapter 6 sheds lights on the challenges and opportunities of real-time network behavior analysis. Chapter 7 presents the benefits and applications of network behavior analysis, while Chap. 8 discusses the research frontiers of network behavior analysis.

References

1. A. Dhamdhere, C. Dovrolis, Ten years in the evolution of the internet ecosystem, in *Proceedings of ACM Internet Measurement Conference (IMC)* (2008)
2. C. Williamson, Internet traffic measurement. IEEE Internet Comput. **5**(6), 70–74 (2001)
3. N. Yaseen, J. Sonchack, V. Liu, Synchronized network snapshots, in *Proceedings of ACM SIGCOMM* (2018)
4. P. Borgnat, G. Dewaele, K. Fukuda, P. Abry, K. Cho, Seven years and one day: sketching the evolution of internet traffic, in *Proceedings of IEEE INFOCOM* (2010)
5. Y. Yuan, D. Lin, A. Mishra, S. Marwaha, R. Alur, B. Loo, Quantitative network monitoring with NetQRE, in *Proceedings of ACM SIGCOMM* (2017)
6. I. Trestian, S. Ranjan, A. Kuzmanovic, A. Nucc, Unconstrained endpoint profiling (googling the internet), in *Proceedings of ACM SIGCOMM* (2008)

7. T. Karagiannis, K. Papagiannaki, N. Taft, M. Faloutsos, Profiling the end host, in *Proceedings of International Conference on Passive and Active Network Measurement* (2007)
8. IPMon, Sprint IP monitoring project, http://ipmon.sprint.com/
9. K. Xu, Z.-L. Zhang, S. Bhattacharyya, Profiling internet backbone traffic: behavior models and applications, in *Proceedings of ACM SIGCOMM* (2005)
10. R. Basat, S. Ramanathan, Y. Li, G. Antichi, M. Yu, M. Mitzenmacher, PINT: probabilistic in-band network telemetry, in *Proceedings of ACM SIGCOMM* (2020)
11. R. Caceres, N. Duffield, A. Feldmann, J.D. Friedmann, A. Greenberg, R. Greer, T. Johnson, C.R. Kalmanek, B. Krishnamurthy, D. Lavelle, P.P. Mishra, J. Rexford, K.K. Ramakrishnan, F.D. True, J.E. van der Memle, Measurement and analysis of IP network usage and behavior. IEEE Commun. Mag. **38**(5), 144–151 (2000)
12. R. Carrasco, M.-A. Sicilia, Unsupervised intrusion detection through skip-gram models of network behavior. Comput. Secur. **78**, 187–197 (2018)
13. T. Johnson, C. Cranor, O. Spatscheck, Gigascope: a stream database for network application, in *Proceedings of ACM SIGMOD* (2003)
14. T. Yang, J. Jiang, P. Liu, Q. Huang, J. Gong, Y. Zhou, R. Miao, X. Li, S. Uhlig, Elastic sketch: adaptive and fast network-wide measurements, in *Proceedings of ACM SIGCOMM* (2018)
15. A. Lakhina, M. Crovella, C. Diot, Diagnosing network-wide traffic anomalies, in *Proceedings of ACM SIGCOMM* (2004)
16. A. Lakhina, M. Crovella, C. Diot, Mining anomalies using traffic feature distributions, in *Proceedings of ACM SIGCOMM* (2005)
17. A. Lazarevic, L. Ertoz, A. Ozgur, J. Srivastava, V. Kumar, A comparative study of anomaly detection schemes in network intrusion detection, in *Proceedings of SIAM International Conference on Data Mining* (2003)
18. MINDS, Minnesota Intrusion Detection System, http://www.cs.umn.edu/research/minds/
19. P. Barford, D. Plonka, Characteristics of network traffic flow anomalies, in *Proceedings of ACM SIGCOMM Internet Measurement Workshop* (2002)
20. P. Barford, J. Kline, D. Plonka, A. Ron, A signal analysis of network traffic anomalies, in *Proceedings of ACM SIGCOMM Internet Measurement Workshop* (2002)
21. A. Rao, Y. Lim, C. Barakat, A. Legout, D. Towsley, W. Dabbous, Network characteristics of video streaming traffic, in *Proceedings of ACM International Conference on emerging Networking EXperiments and Technologies (CoNEXT)* (2011)
22. S. Sen, J. Wang, Analyzing peer-to-peer traffic across large networks. IEEE/ACM Trans. Netw. 219–232 (2004)
23. T. Callahan, M. Allman, V. Paxson, A longitudinal view of HTTP traffic, in *Proceedings of International Conference on Passive and Active Network Measurement* (2010)
24. T. Karagiannis, A. Broido, M. Faloutsos, K.C. Claffy, Transport layer identification of P2P traffic, in *Proceedings of ACM Internet Measurement Conference* (2004)
25. G. Gürsun, N. Ruchansky, E. Terzi, M. Crovella, Inferring visibility: who's (not) talking to whom?, in *Proceedings of ACM SIGCOMM* (2012)
26. K. Xu, F. Wang, Behavioral graph analysis of internet applications, in *Proceedings of IEEE GLOBECOM* (2011)
27. K. Xu, F. Wang, L. Gu, Network-aware behavior clustering of internet end hosts, in *Proceedings of IEEE INFOCOM* (2011)
28. K. Xu, F. Wang, L. Gu, Behavior analysis of internet traffic via bipartite graphs and one-mode projection. IEEE/ACM Trans. Netw. **22**(3), 931–942 (2014)
29. K. Xu, F. Wang, S. Bhattacharyya, Z.-L. Zhang, A real-time network traffic profiling system, in *Proceedings of International Conference on Dependable Systems and Networks* (2007)
30. K. Xu, F. Wang, S. Bhattacharyya, Z.-L. Zhang, Real-time behavior profiling for network monitoring. Int. J. Internet Protoc. Technol. **5**(1/2), 65–80 (2010)
31. K. Xu, Z.-L. Zhang, S. Bhattacharyya, Reducing unwanted traffic in a backbone network, in *Proceedings of Steps to Reducing Unwanted Traffic on the Internet Workshop (SRUTI)* (2005)
32. K. Xu, Z.-L. Zhang, S. Bhattacharyya, Internet traffic behavior profiling for network security monitoring. IEEE/ACM Trans. Netw. **16**, 1241–1252 (2008)

33. X. Wei, N. Valler, H. Madhyastha, I. Neamtiu, M. Faloutsos, A behavior-aware profiling of handheld devices, in *Proceedings of IEEE INFOCOM* (2015)
34. A. Lakhina, K. Papagiannaki, M. Crovella, C. Diot, E. Kolaczyk, N. Taft, Structural analysis of network traffic flows, in *Proceedings of ACM SIGMETRICS* (2004)
35. D. Figueiredo, B. Liu, V. Misra, D. Towsley, On the autocorrelation structure of TCP traffic. Comput. Netw. **40**(3), 339–361 (2002)
36. W. Leland, M. Taqqu, W. Willinger, D. Wilson, On the self-similar nature of ethernet traffic. IEEE/ACM Trans. Netw. **2**(1), 1–15 (1994)
37. M. Iliofotou, M. Faloutsos, M. Mitzenmacher, Exploiting dynamicity in graph-based traffic analysis: techniques and applications, in *Proceedings of ACM International Conference on emerging Networking EXperiments and Technologies (CoNEXT)* (2009)
38. M. Iliofotou, P. Pappu, M. Faloutsos, M. Mitzenmacher, S. Singh, G. Varghese, Network monitoring using traffic dispersion graphs (TDGs), in *Proceedings of ACM Internet Measurement Conference (IMC)* (2007)
39. S. Xue, L. Zhang, A. Li, X.-Y. Li, C. Ruan, W. Huang, AppDNA: app behavior profiling via graph-based deep learning, in *Proceedings of IEEE INFOCOM* (2018)
40. H. Kim, K. Claffy, M. Fomenkov, D. Barman, M. Faloutsos, K. Lee, Internet traffic classification demystified: myths, caveats, and the best practices, in *Proceedings of ACM International Conference on emerging Networking EXperiments and Technologies (CoNEXT)* (2008)
41. M. Dischinger, M. Marcon, S. Guha, K. Gummadi, R. Mahajan, S. Saroiu, Glasnost: enabling end users to detect traffic differentiation, in *Proceedings of USENIX Symposium on Networked Systems Design and Implementation (NSDI)* (2010)
42. M. Tai, S. Ata, I. Oka, Fast, accurate, and lightweight real-time traffic identification method based on flow statistics, in *Proceedings of International Conference on Passive and Active Network Measurement* (2007)
43. J. Tang, Y. Cheng, Selfish misbehavior detection in 802.11 based wireless networks: an adaptive approach based on Markov decision process, in *Proceedings of IEEE INFOCOM* (2013)
44. E. Raftopoulos, X. Dimitropoulos, Detecting, validating and characterizing computer infections in the wild, in *Proceedings of ACM Internet Measurement Conference (IMC)* (2011)
45. M. Vallentin, V. Paxson, R. Sommer, VAST: a unified platform for interactive network forensics, in *Proceedings of USENIX Symposium on Networked Systems Design and Implementation (NSDI)* (2016)
46. S. Zander, N. Williams, G. Armitage, Internet archeology: estimating individual application trends in incomplete historic traffic traces, in *Proceedings of International Conference on Passive and Active Network Measurement* (2006)
47. A. Sarabi, M. Liu, Characterizing the internet host population using deep learning: a universal and lightweight numerical embedding, in *Proceedings of ACM Internet Measurement Conference (IMC)* (2018)
48. U. Weinsberg, A. Soule, L. Massoulie, Inferring traffic shaping and policy parameters using end host measurements, in *Proceedings of IEEE INFOCOM* (2011)
49. R. Gonzalez, C. Soriente, N. Laoutaris, User profiling in the time of HTTPS, in *Proceedings of ACM Internet Measurement Conference (IMC)* (2016)
50. T. Qiu, Z. Ge, S. Lee, J. Wang, J. Xu, Q. Zhao, Modeling user activities in a large IPTV system, in *Proceedings of ACM Internet Measurement Conference (IMC)* (2009)
51. Z. Bischof, F. Bustamante, R. Stanojevic, Need, want, can afford – broadband markets and the behavior of users, in *Proceedings of ACM Internet Measurement Conference (IMC)* (2014)
52. E. Liang, H. Zhu, X. Jin, I. Stoica, Neural packet classification, in *Proceedings of ACM SIGCOMM* (2019)
53. F. Tegeler, X. Fu, G. Vigna, C. Kruegel, BotFinder: finding bots in network traffic without deep packet inspection, in *Proceedings of ACM International Conference on emerging Networking EXperiments and Technologies (CoNEXT)* (2012)
54. R. Perdisci, W. Lee, N. Feamster, Behavioral clustering of HTTP-based malware and signature generation using malicious network traces, in *Proceedings of USENIX Symposium on Networked Systems Design and Implementation (NSDI)* (2010)

Chapter 2
Background of Network Behavior Analysis

Abstract Network behavior analysis, a research subfield of Internet measurement and analysis, is centered on the collection and analysis of network traffic data for unveiling behavioral patterns and communication structure of networked systems and Internet applications. This chapter first provides a brief background on Internet measurement and analysis and subsequently explains the tools, instruments, and facility of data collection for network behavior analysis. Finally, this chapter sheds on the basic and advanced analysis of network traffic features via entropy measures in information theory, bipartite graphs, and one-mode projections. The basic analysis of traffic features is the early and critical step for exploring advanced analysis of network traffic and for developing behavioral, structural, and graphical models of network behavior for networked systems and Internet applications.

2.1 Internet Measurement and Analysis

The last few decades have witnessed the growing impact of the Internet on the society thanks to the continuous innovations of Internet devices such as smartphones and Internet of things and application services such as World Wild Web, email, social media and networks, video streaming, virtual meetings, and online education. The applications have fundamentally changed the communication, entertainment, commerce, media, and culture and have played a key role in rapidly transitioning billions of people to remote working and learning from home during the coronavirus pandemics [3]. The critical importance of the Internet calls for a deep understanding on how the Internet works and behaves via network measurement, monitoring and analysis, and leads to the rapid development and advance of the emerging research field in Internet measurement and analysis.

The primary goal of Internet measurement and analysis is to provide insights on how the Internet works and behaves via monitoring, measuring, collecting, analyzing, and modeling the Internet from a variety of perspectives including traffic, routing, applications, performance, and security [25, 39, 48, 68, 77]. The areas of

K. Xu, *Network Behavior Analysis*,
https://doi.org/10.1007/978-981-16-8325-1_2

Internet measurement and analysis include Internet topology [16, 33, 46, 91], network performance [4, 20, 21, 27, 40, 82], traffic analysis [6, 51, 58, 60, 84, 88, 94], Internet routing [29, 36], network security and forensics [64, 66, 96], diagnosis, debugging, and troubleshooting [2, 9, 65, 90, 93], and Internet monitoring [7, 34, 61, 83, 85]. The networks of interests are very broad, covering IP backbone and transit networks [12, 53, 57, 71], peer-to-peer networks [23, 24, 86], overlay networks [18, 97], content distribution networks [49, 70, 72], enterprise networks [50, 75], data center networks [8, 63, 76, 89], as well as edge networks [15, 19, 44, 69, 78, 95] such as home networks [30, 42, 43, 52, 74], vehicle networks [13, 92], and cellular networks [32, 47, 56].

A rich body of research literature [37, 73, 79, 80, 87] has studied Internet characteristics on topology, structure, workload, application, and end users in wired or wireless networks. The topology and structure characterizations of the Internet attempt to map the Internet infrastructure at a variety of coarse and fine granularities including networked systems, IP addresses, routers, autonomous systems (ASNs), domains, and Internet service providers (ISPs). These topology maps and Internet connectivity analysis provide critical insights on end-to-end performance, bandwidth optimization and capacity planning, and the ASN relationship inference between neighboring ISPs.

The research on Internet traffic analysis [41, 55, 98] has extensively studied statistical patterns and properties in network traffic from IP backbone networks, edge networks, enterprise networks, data center networks, and wireless networks. The Internet traffic analysis gives insights on traffic classification and characteristics, statistical properties, workload composition, and network traffic evolution. Given the "big data" nature of Internet traffic, many researchers have developed effective traffic sketching, summarizing, and streaming algorithms for measuring, characterizing, and modeling network traffic.

Internet routing is responsible for forwarding IP data packets from the original end system to the destination system over the Internet. To support the massive scale of the Internet with billions of end systems, the routing protocols of the Internet forms a two-layered hierarchical structure: inter-domain routing and intra-domain routing. The inter-domain routing protocol handles the exchange, announcement, and withdrawal of routing information between different ASNs, while the intra-domain routing protocol exchanges on the routing information for the routers within the single ASN under the same network administration and management. The de facto inter-domain routing protocol is BGP (Border Gateway Protocol), while the intra-domain routing protocol is independently chosen from different protocols such as OSPF (Open Shortest Path First), RIP (Routing Information Protocol), and IGRP (Interior Gateway Routing Protocol) by the respective network engineering team at different ISPs. As the fundamental infrastructure of the Internet, Internet routing has been extensively studied in the literature, particularly on the policies, performance, stability, convergence, security, and misconfigurations.

Due to the prevalent and continuous threats and cyberattacks towards networked systems and Internet infrastructure, a significant number of the research efforts on Internet measurement and analysis is devoted to network security and digital forensics. These measurement-based approaches collect and analyze traffic and routing data from networked systems and devices to mine anomalous patterns and emerging zero-day behaviors for anomaly detection, threat modeling, and forensic analysis.

Internet monitoring is a crucial component for analyzing how the Internet works, behaves, and most importantly evolves over time. Towards establishing scalable Internet monitoring platforms for collecting long-term, trustworthy, and representative Internet measurement, the networking and system research communities have developed several planet-scale platforms and testbeds, designed a suite of new traffic sampling and sketching strategies, and introduced privacy-preserving anonymization methods for data sharing.

The research topic of network behavior analysis is at the intersections of Internet characterization, traffic analysis, network security and forensics, and Internet monitoring. Network behavior analysis shares data collection techniques, analysis methodologies, and monitoring strategies with Internet characterization, traffic analysis, and Internet monitoring. Network behavior analysis has a wide range of mission-critical applications for network operation and management, including network security and digital forensics

2.2 Data Collection for Network Behavior Analysis

Network behavior analysis relies on raw or processed Internet traffic data such as IP data packets and network flow records for making sense of the underlying behaviors of networked systems and Internet applications. Figure 2.1 illustrates the end-to-end process of network behavior analysis. The process starts with network traffic data collections, data storage and preprocessing, and continues with network behavioral analysis and modeling after behavioral feature selection and explorations. Finally, the process produces actionable behavioral insights for various applications. Thus, collecting, storing, and archiving traffic data is a crucial and early component of network behavior analysis and its applications [26].

Internet measurement is typically performed in one of two different infrastructures: active measurement and passive measurement [28]. The active measurement infrastructure creates carefully crafted probing packets into the networks for specific measurement tasks such as quantifying end-to-end latency, measuring available network bandwidth, and discovering the Internet topology [38, 54]. Different from active measurement, the passive measurement infrastructure strategically selects vantage points such as network routers in ISPs and broadband home routers in residential networks for measuring and monitoring various network characteristics and properties [1, 10]. In other words, the key difference between active measurement and passive measurement infrastructures lies in the creation of new data packets into the networks under measurement.

Data collection for network analysis often relies on passive measurement infrastructures to passively monitor and collect traffic data at network routers and end systems. The temporal frequency of such data collection can be continuous, periodical,

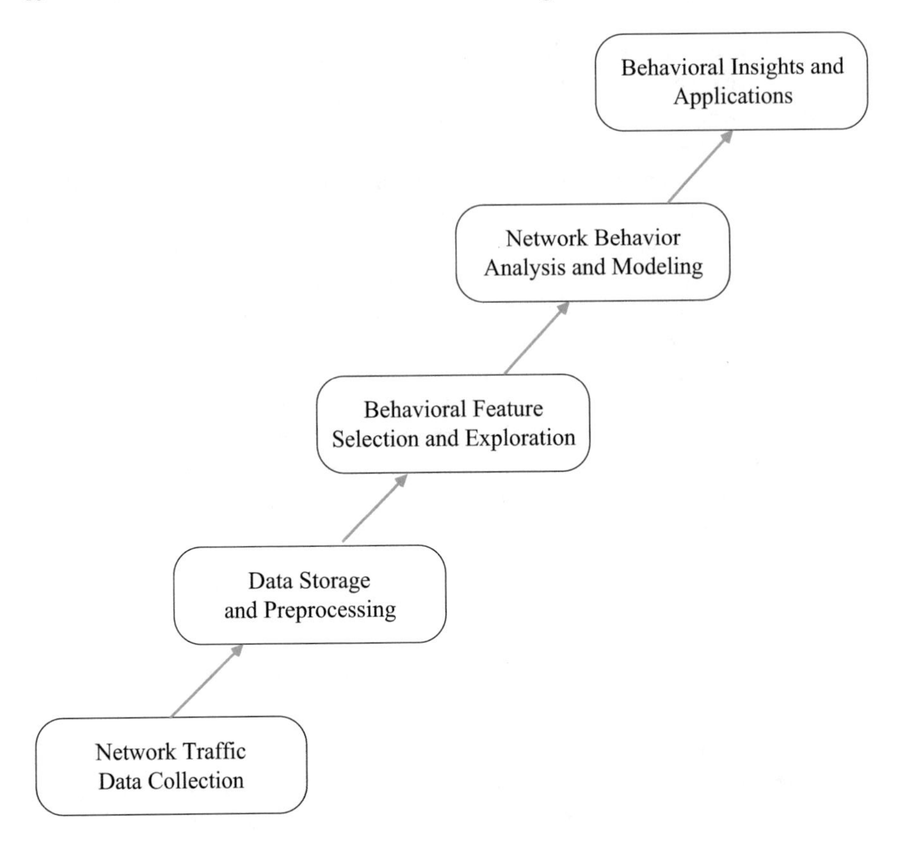

Fig. 2.1 The end-to-end process of network behavior analysis

or on-demand. For example, the network telescope project at University of California San Diego (UCSD) builds an always-on and passive monitoring platform [22] based on a lightly utilized class A network, represented as /8 in classless inter-domain routing (CIDR) notation, to continuously monitor and collect unproductive or unwanted data traffic towards networked systems and the very large unused IP address space in the network. The traffic data from the network telescope provides insights on the prevalence and intensity of anomaly traffic on the Internet due to self-replicating and propagating computer worms, Internet backscatters, vulnerability scanning by botnets, and many other Internet background radiation events.

The collected data for network behavior analysis has two major types: raw IP data packets and aggregated network flows [99]. The benefit of collecting raw IP data packets lies in the availability of all features in data communications. However, collecting and storing raw IP traffic has an inherent challenge of storing and archiving the sheer volume of data, e.g., Netflix has estimated 4.5 GB traffic data for the streaming or downloading of a single 90-minute high-definition (HD) movie. In the practice of raw IP data packet collections, some systems choose to collect the entire

IP packet including both the packet header and the full payload, while the other systems, which are mostly interested in the features in the packet headers and the first few bytes in the payload, choose to only collect the first 40 bytes of the IP packets for reducing unnecessary data storage and processing [14].

The network flow data, summarizing IP data packets in data communications between end systems, requires far less data for each network conversation. For example, the same video streaming session, if collected in the form of network flows, costs only hundreds of bytes, thus saving data storage and processing for several orders of magnitude. On the other hand, network flow data lacks the complete picture and traffic features of data communications between end systems, e.g., inter-packet arrival times and the distributions of IP data packet sizes in the same data communication.

In addition to raw IP data packets and network flow records, networking researchers also collect auxiliary datasets for network behavior analysis. The auxiliary datasets include Internet routing tables for characterizing network prefix origins and best route selections by ISPs, Internet exchange points (IXPs) peering database for understanding where ISPs and CDNs exchange network traffic, and IP geolocation data for mapping the geographical locations of IP addresses and network prefixes [59, 62, 67].

2.3 Preliminaries of Network Behavior Analysis

2.3.1 Information Theory and Entropy

In the literature, network traffic analysis relies on volume-based and distribution-based approaches to study traffic features. The volume-based approach focuses on the simple counting on the observations on the traffic features, and often provides many valuable summaries on network traffic, e.g., how many networked systems does a smartphone communicate with or how many outgoing and incoming IP data packet counts and byte counts does a web server send and receive during a 5-minute time window. However, the volume-based approach lacks the ability to shed light on the variations, distributions, or patterns inside the absolute volumes. For example, two web servers, receiving the equal amount of one million IP data packets during the same time period, might exhibit dramatically different behavioral patterns. One server might communicate with thousands of random web browsers across the Internet which collectively send one million IP data packets, while another server might be under distributed denial-of-service (DDoS) attacks from exactly one million unique source IP addresses each of which sends one single TCP SYN segment. Therefore, the networking research community has developed distribution-based approaches for effectively distinguishing such different patterns under the same traffic volumes.

2.3.1.1 Entropy Measures

To complement the volume-based approach on traffic feature analysis, several research studies [5, 11, 31, 45] have introduced the distribution-based approach to characterize the distributions of traffic features via probability and entropy concepts from probability theory and information theory. Information essentially quantifies "the amount of uncertainty" contained in data [81]. Consider a random variable X that may take N_X discrete values. Suppose we randomly sample or observe X for m times, which induces an empirical probability distribution on X, $p(x_i) = m_i/m, x_i \in X$, where m_i is the frequency or number of times we observe X taking the value x_i. The (empirical) *entropy* of X is then defined as

$$H(X) := -\sum_{x_i \in X} p(x_i) \log p(x_i), \tag{2.1}$$

where by convention $0 \log 0 = 0$.

Entropy measures the "observational variety" in the observed values of X [17]. Note that unobserved possibilities (due to $0 \log 0 = 0$) do not enter the measure, and $0 \leq H(X) \leq H_{max}(X) := \log \min\{N_X, m\}$. $H_{max}(X)$ is often referred to as the *maximum entropy* of (sampled) X, as $2^{H_{max}(X)}$ is the maximum number of possible *unique* values (i.e., "maximum uncertainty") that the observed X can take in m observations.

2.3.1.2 Standardized Entropy and Relative Uncertainty

Clearly, the entropy measure $H(X)$ is a function of the support size N_X and sample size m. Assuming that $m \geq 2$ and $N_X \geq 2$ (otherwise there is no "observational variety" to speak of), we define the *standardized* entropy below—referred to as *relative uncertainty* (RU), as it provides an index of variety or uniformity regardless of the support or sample size:

$$RU(X) := \frac{H(X)}{H_{max}(X)} = \frac{H(X)}{\log \min\{N_X, m\}}. \tag{2.2}$$

Clearly, if $RU(X) = 0$, then all observations of X are of the same kind, i.e., $p(x) = 1$ for some $x \in X$; thus observational variety is completely absent. More generally, let A denote the (sub)set of observed values in X, i.e., $p(x_i) > 0$ for $x_i \in A$. Suppose $m \leq N_X$. Then $RU(X) = 1$ if and only if $|A| = m$ and $p(x_i) = 1/m$ for each $x_i \in A$. In other words, all observed values of X are different or *unique*, thus the observations have the highest degree of variety or uncertainty. Hence, when $m \leq N_X$, $RU(X)$ provides a measure of "randomness" or "uniqueness" of the values that the observed X may take—this is what is mostly used in network traffic analysis, as in general $m \ll N_X$.

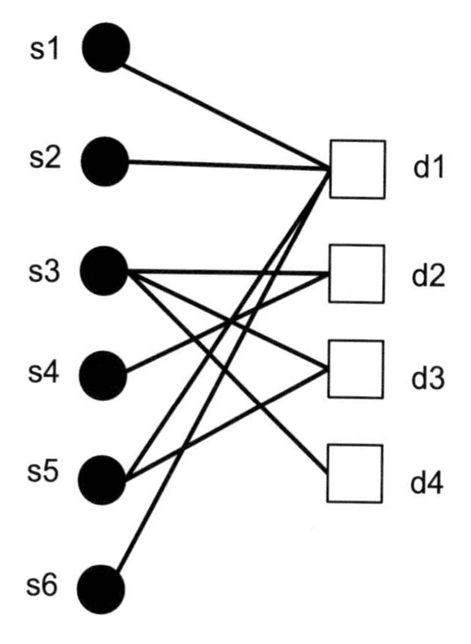

Fig. 2.2 Modeling host communications using bipartite graphs

In the case of $m > N_X$, $RU(X) = 1$ if and only if $m_i = m/N_X$, thus $p(x_i) = 1/N_X$ for $x_i \in A = X$, i.e., the observed values are *uniformly* distributed over X. In this case, $RU(X)$ measures the degree of uniformity in the observed values of X. As a general measure of *uniformity* in the observed values of X, we consider the conditional entropy $H(X|A)$ and *conditional relative uncertainty* $RU(X|A)$ by conditioning X based on A. Then we have $H(X|A) = H(X)$, $H_{max}(X|A) = \log|A|$, and $RU(X|A) = H(X)/log|A|$. Hence, $RU(X|A) = 1$ if and only if $p(x_i) = 1/|A|$ for every $x_i \in A$. In general, $RU(X|A) \approx 1$ means that the observed values of X are closer to being uniformly distributed, thus less distinguishable from each other, whereas $RU(X|A) \ll 1$ indicates that the distribution is more skewed, with a few values more frequently observed.

2.3.2 *Graphical Analysis*

2.3.2.1 Bipartite Graphs of Host Communications

Host communications observed in network traffic of Internet links could be naturally modeled with a bipartite graph $\mathcal{G} = (\mathcal{A}, \mathcal{B}, \mathcal{E})$, where $\mathcal{A}$ and $\mathcal{B}$ are two disjoint vertex sets, and $\mathcal{E} \subseteq \mathcal{A} \times \mathcal{B}$ is the edge set [35]. Specifically, all the `source` IP addresses observed in network traffic from one single direction of an Internet link form the vertex set $\mathcal{A}$, while the vertex set $\mathcal{B}$ consists of all the `destination` addresses observed in the same traffic. Each of the edges, e_k in $\mathcal{G}$ connects one vertex $a_i \in \mathcal{A}$ and another vertex $b_j \in \mathcal{B}$.

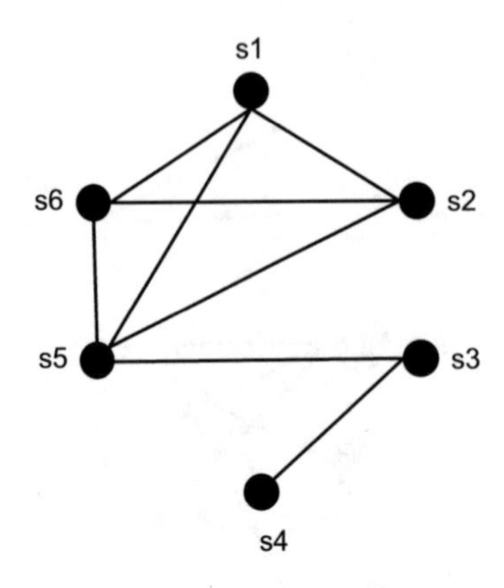

(a) One-mode projection graph of source hosts

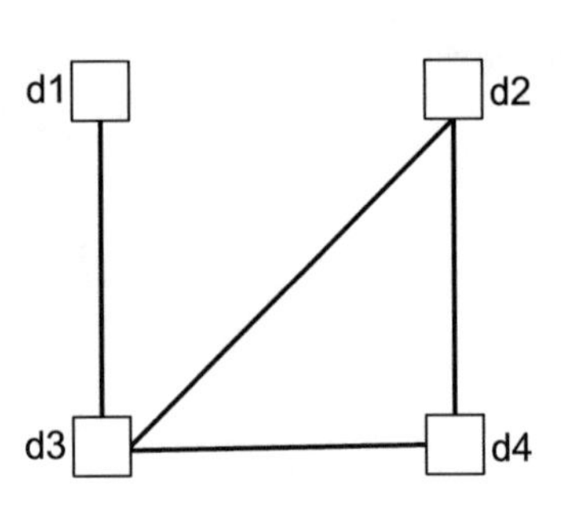

(b) One-mode projection graph of destination hosts

Fig. 2.3 Modeling the social-behavior similarity of networked systems with one-mode projection of bipartite graphs

Figure 2.2 illustrates an example of a simple bipartite graph that shows data communications between six source IP addresses (s_1 - s_6) and four destination IP addresses (d_1 - d_4). Note that an Internet link carries network traffic from two directions, thus we separate network traffic based on traffic directions and use bipartite graphs to model network traffic from two directions separately.

2.3.2.2 One-Mode Projections of Bipartite Graphs

To study the social-behavior similarity of end hosts in network traffic, we leverage one-mode projection graphs of bipartite graphs that are used to extract hidden information or relationships between nodes within the same vertex sets [35]. Figure 2.3[a] illustrates the one-mode projection of the bipartite graph on the vertex set of the six left-side nodes, i.e., the source hosts (s_1 - s_6) in Fig. 2.2, while Fig. 2.3[b] is the one-mode projection on the four destination hosts d_1 - d_4 in Fig. 2.2. An edge connects two nodes in the one-mode projection if and only if both nodes have connections to at least one same node in the bipartite graph. Thus studying one-mode projection graphs could potentially reveal the similarity or dissimilarity of communication patterns and traffic behaviors for networked systems and Internet applications.

References

1. A. Dhamdhere, L. Breslau, N. Duffield, C. Ee, A. Gerber, C. Lund, S. Sen, FlowRoute: inferring forwarding table updates using passive flow-level measurements, in *Proceedings of ACM Internet Measurement Conference (IMC)* (2010)
2. A. Dhamdhere, R. Teixeira, C. Dovrolis, C. Diot, NetDiagnoser: troubleshooting network unreachabilities using end-to-end probes and routing data, in *Proceedings of ACM International Conference on emerging Networking EXperiments and Technologies (CoNEXT)* (2007)
3. A. Feldmann, O. Gasser, F. Lichtblau, E. Pujol, I. Poese, C. Dietzel, D. Wagner, M. Wichtlhuber, J. Tapiador, N. Vallina-Rodriguez, O. Hohlfeld, G. Smaragdakis, A year in lockdown: how the waves of COVID-19 impact internet traffic. Commun. ACM **64**(7), 101–108 (2021)
4. A. Jaggard, S. Kopparty, V. Ramachandran, R. Wright, The design space of probing algorithms for network-performance measurement, in *Proceedings of ACM SIGMETRICS* (2013)
5. A. Lakhina, M. Crovella, C. Diot, Mining anomalies using traffic feature distributions, in *Proceedings of ACM SIGCOMM* (2005)
6. A. Lall, V. Sekar, M. Ogihara, J. Xu, H. Zhang, Data streaming algorithms for estimating entropy of network traffic, in *Proceedings of ACM SIGMETRICS* (2006)
7. A. Pietro, F. Huici, D. Costantini, S. Niccolini, DECON: decentralized coordination for large-scale flow monitoring, in *Proceedings of IEEE INFOCOM* (2010)
8. A. Roy, D. Bansal, D. Brumley, H. Chandrappa, P. Sharma, R. Tewari, B. Arzani, A. Snoeren, Cloud datacenter SDN monitoring: experiences and challenges, in *Proceedings of ACM Internet Measurement Conference (IMC)* (2018)
9. B. Aggarwal, R. Bhagwan, L. Carli, V. Padmanabhan, K. Puttaswamy, Deja vu: fingerprinting network problems, in *Proceedings of ACM International Conference on emerging Networking EXperiments and Technologies (CoNEXT)* (2011)
10. B. Eriksson, P. Barford, R. Nowak, M. Crovella, Learning network structure from passive measurements, in *Proceedings of ACM Internet Measurement Conference (IMC)* (2007)
11. B. Tellenbach, M. Burkhart, D. Sornette, T. Maillart, Beyond shannon: characterizing internet traffic with generalized entropy metrics, in *Proceedings of International Conference on Passive and Active Network Measurement* (2009)
12. B.-Y. Choi, S. Moon, R. Cruz, Z.-L. Zhang, C. Diot, Practical delay monitoring for ISPs, in *Proceedings of ACM International Conference on emerging Networking EXperiments and Technologies (CoNEXT)* (2005)
13. C. Andrade, S. Byers, V. Gopalakrishnan, E. Halepovic, D. Poole, L. Tran, C. Volinsky, Connected cars in cellular network: a measurement study, in *Proceedings of ACM Internet Measurement Conference (IMC)* (2017)
14. C. Fraleigh, S. Moon, B. Lyles, C. Cotton, M. Khan, D. Moll, R. Rockell, T. Seely, and S. Diot: Packet-Level Traffic Measurements from the Sprint IP Backbone. IEEE Network **17**(6) (2003)
15. C. Kreibich, N. Weaver, B. Nechaev, and V. Paxson: Netalyzr: Illuminating The Edge Network. In: Proceedings of ACM Internet Measurement Conference (IMC) (2010)
16. C. Liu, A. Swami, D. Towsley, T. Salonidis, A. Bejan, and P. Yu: Fisher Information-based Experiment Design for Network Tomography. In: Proceedings of ACM SIGMETRICS (2015)
17. Cover, T., Thomas, J.: Elements of Information Theory. Wiley Series in Telecommunications (1991)
18. D. Andersen, H. Balakrishnan, F. Kaashoek, and R. Morris: Resilient Overlay Networks. In: Proceedings of ACM Symposium on Operating Systems Principles (SOSP) (2001)
19. D. Croce, T. En-Najjary, G. Urvoy-Keller, and E. Biersack: Capacity Estimation of ADSL links. In: Proceedings of ACM International Conference on emerging Networking EXperiments and Technologies (CoNEXT) (2008)
20. D. Lee, K. Jang, C. Lee, G. Iannaccone, and S. Moon: Path Stitching: Internet-Wide Path and Delay Estimation from Existing Measurements. In: Proceedings of IEEE INFOCOM (2010)
21. D. Leonard and D. Loguinov: Turbo King: Framework for Large-Scale Internet Delay Measurements. In: Proceedings of IEEE INFOCOM (2008)

22. D. Moore, V. Paxson, S. Savage, C. Shannon, S. Staniford, and N. Weaver: Inside the Slammer Worm. IEEE Security and Privacy (2003)
23. D. Stutzbach and R. Rejaie: Understanding Churn in Peer-to-Peer Networks. In: Proceedings of ACM Internet Measurement Conference (IMC) (2006)
24. D. Stutzbach, R. Rejaie, N. Duffield, S. Sen, and W. Willinger: On Unbiased Sampling for Unstructured Peer-to-Peer Networks. In: Proceedings of ACM Internet Measurement Conference (IMC) (2006)
25. F. Baccelli, S. Machiraju, D. Veitch, and J. Bolot: The Role of PASTA in Network Measurement. In: Proceedings of ACM SIGCOMM (2006)
26. F. Fusco, M. Vlachos, and X. Dimitropoulos: RasterZip: Compressing Network Monitoring Data with Support for Partial Decompression. In: Proceedings of ACM Internet Measurement Conference (IMC) (2012)
27. F. Uyeda, L. Foschini, F. Baker, S. Suri, and G. Varghese: Efficiently Measuring Bandwidth at All Time Scales. In: Proceedings of USENIX Symposium on Networked Systems Design and Implementation (NSDI) (2011)
28. G. Bartlett, J. Heidemann, and C. Papadopoulos: Understanding Passive and Active Service Discovery. In: Proceedings of ACM Internet Measurement Conference (IMC) (2007)
29. G. Comarela, G. Gursun, and M. Crovella: Studying Interdomain Routing over Long Timescales. In: Proceedings of ACM Internet Measurement Conference (IMC) (2013)
30. G. Maier, A. Feldmann, V. Paxson, and M. Allman: On Dominant Characteristics of Residential Broadband Internet Traffic. In: Proceedings of ACM Internet Measurement Conference (IMC) (2009)
31. G. Nychis, V. Sekar, D. Andersen, H. Kim, and H. Zhang: An Empirical Evaluation of Entropy-based Traffic Anomaly Detection. In: Proceedings of ACM Internet Measurement Conference (IMC) (2008)
32. H. Deng, C. Peng, A. Fida, J. Meng, and Y. Hu: Mobility Support in Cellular Networks: A Measurement Study on Its Configurations and Implications. In: Proceedings of ACM Internet Measurement Conference (IMC) (2018)
33. H. Li, Y. Gao, W. Dong, C. Chen, Preferential Link Tomography in Dynamic Networks. IEEE/ACM Transactions on Networking **27**(5), 1801–1814 (2019)
34. H. Song, L. Qiu, Y. Zhang, NetQuest: A Flexible Framework for Large-Scale Network Measurement. IEEE/ACM Transactions on Networking **17**(1), 106–119 (2009)
35. J.-L. Guillaume, M. Latapy, Bipartite graphs as models of complex networks. Physica A: Statistical and Theoretical Physics **371**(2), 795–813 (2006)
36. J. Ni, H. Xie, S. Tatikonda, and Y. Yang: Network Routing Topology Inference from End-to-End Measurements. In: Proceedings of IEEE INFOCOM (2008)
37. J. Rasley, B. Stephens, C. Dixon, E. Rozner, W. Felter, K. Agarwal, J. Carter, and R. Fonseca: Planck: Millisecond-Scale Monitoring and Control for Commodity Networks. In: Proceedings of ACM SIGCOMM (2014)
38. J. Sommers and P. Barford: An Active Measurement System for Shared Environments. In: Proceedings of ACM Internet Measurement Conference (IMC) (2007)
39. J. Sommers, P. Barford, N. Duffield, and A. Ron: Accurate and Efficient SLA Compliance Monitoring. In: Proceedings of ACM SIGCOMM (2007)
40. J. Wang, S. Lian, W. Dong, Y. Liu, and X.-Y. Li: Every Packet Counts: Fine-Grained Delay and Loss Measurement with Reordering. In: Proceedings of IEEE International Conference on Network Protocols (ICNP) (2014)
41. K. Papagiannaki, N. Taft, Z.-L. Zhang, and C. Diot: Long-Term Forecasting of Internet Backbone Traffic: Observations and Initial Models. In: Proceedings of IEEE INFOCOM (2003)
42. K. Xu, F. Wang, X. Jia, Secure the Internet, One Home at a Time. Security and Communication Networks **9**(16), 3821–3832 (2016)
43. K. Xu, F. Wang, L. Gu, J. Gao, Y. Jin, Characterizing Home Network Traffic: An Inside View. Accepted by Personal and Ubiquitous Computing **18**(4), 967–975 (2014)
44. K. Xu, Y. Wan, G. Xue, and F. Wang: Multidimensional Behavioral Profiling of Internet-of-Things in Edge Networks. In: Proceedings of IEEE/ACM International Symposium on Quality of Service (IWQoS) (2019)

45. K. Xu, Z.-L. Zhang, and S. Bhattacharyya: Profiling Internet Backbone Traffic: Behavior Models and Applications. In: Proceedings of ACM SIGCOMM (2005)
46. L. Ma, T. He, K. Leung, A. Swami, and D. Towsley: Monitor Placement for Maximal Identifiability in Network Tomography. In: Proceedings of IEEE INFOCOM (2014)
47. L. Xue, X. Ma, X. Luo, L. Yu, S. Wang, and T. Chen: Is What You Measure What You Expect? Factors Affecting Smartphone-Based Mobile Network Measurement. In: Proceedings of IEEE INFOCOM (2017)
48. L. Yuan, C.-N. Chuah, and P. Mohapatra: ProgME: towards Programmable Network Measurement. In: Proceedings of ACM SIGCOMM (2007)
49. M. Calder, A. Flavel, E. Katz-Bassett, R. Mahajan, and J. Padhye: Analyzing the Performance of an Anycast CDN. In: Proceedings of ACM Internet Measurement Conference (IMC) (2015)
50. M. Casado, T. Garfinkel, A. Akella, M. Freedman, D. Boneh, N. McKeown, and S. Shenker: SANE: A Protection Architecture for Enterprise Networks. In: Proceedings of USENIX Security Symposium (2006)
51. M. Chen, S. Chen, Z. Cai, Counter Tree: A Scalable Counter Architecture for Per-Flow Traffic Measurement. IEEE/ACM Transactions on Networking **25**(2), 1249–1262 (2017)
52. M. Dischinger, A. Haeberlen, K. Gummadi, and S. Saroiu: Characterizing Residential Broadband Networks. In: Proceedings of ACM Internet Measurement Conference (IMC) (2007)
53. M. Iliofotou, B. Gallagher, T. Eliassi-Rad, G. Xie, and M. Faloutsos: Profiling-by-Association: A Resilient Traffic Profiling Solution for the Internet Backbone. In: Proceedings of ACM International Conference on emerging Networking EXperiments and Technologies (CoNEXT) (2010)
54. M. Luckie: Scamper: a Scalable and Extensible Packet Prober for Active Measurement of the Internet. In: Proceedings of ACM Internet Measurement Conference (IMC) (2010)
55. M. Roughan, A. Greenberg, C. Kalmanek, M. Rumsewicz, J. Yates, and Y. Zhang: Experience in Measuring Backbone Traffic Variability: Models, Metrics, Measurements and Meaning. In: Proceedings of ACM Internet Measurement Workshop (2002)
56. M. Shafiq, L. Ji, A Liu, and J. Wang: Characterizing and Modeling Internet Traffic Dynamics of Cellular Devices. In: Proceedings of ACM SIGMETRICS (2011)
57. M. Trevisan, D. Giordano, I. Drago, M. Mellia, and M. Munafo: Five Years at the Edge: Watching Internet from the ISP Network. In: Proceedings of ACM International Conference on emerging Networking EXperiments and Technologies (CoNEXT) (2018)
58. M. Yu, L. Jose, and R. Miao: Software Defined Trafic Measurement with OpenSketch. In: Proceedings of USENIX Symposium on Networked Systems Design and Implementation (NSDI) (2013)
59. MaxMind: Geoip databases. https://www.maxmind.com/en/geoip2-databases
60. N. Duffield: Fair Sampling across Network Flow Measurements. In: Proceedings of ACM SIGMETRICS (2012)
61. O. Argon, Y. Shavitt, and U. Weinsberg: Inferring the Periodicity in Large-Scale Internet Measurements. In: Proceedings of IEEE INFOCOM (2013)
62. of Oregon, U.: Routeviews archive project. http://archive.routeviews.org/
63. P. Gill, N. Jain, and N. Nagappan: Understanding Network Failures in Data Centers: Measurement, Analysis, and Implications. In: Proceedings of ACM SIGCOMM (2011)
64. P. Richter and A. Berger: Scanning the Scanners: Sensing the Internet from a Massively Distributed Network Telescope. In: Proceedings of ACM Internet Measurement Conference (IMC) (2019)
65. P. Tammana, R. Agarwal, and M. Lee: Distributed Network Monitoring and Debugging with SwitchPointer. In: Proceedings of USENIX Symposium on Networked Systems Design and Implementation (NSDI) (2018)
66. P. Wang, P. Jia, J. Tao, and X. Guan: Mining Long-Term Stealthy User Behaviors on High Speed Links. In: Proceedings of IEEE INFOCOM (2018)
67. PeeringDB: The interconnection database. https://www.peeringdb.com/
68. Q. Huang, X. Jin, P. Lee, R. Li, L. Tang, Y. Chen, and G. Zhang: SketchVisor: Robust Network Measurement for Software Packet Processing. In: Proceedings of ACM SIGCOMM (2017)

69. R. Fontugne, A. Shah, and K. Cho: Persistent Last-mile Congestion: Not so Uncommon. In: Proceedings of ACM Internet Measurement Conference (IMC) (2020)
70. R. Krishnan, H. Madhyastha, S. Srinivasan, S. Jain, A. Krishnamurthy, T. Anderson, and J. Gao: Moving Beyond End-to-End Path Information to Optimize CDN Performance. In: Proceedings of ACM Internet Measurement Conference (IMC) (2009)
71. R. Mahajan, M. Zhang, L. Poole, and V. Pai: Uncovering Performance Differences in Backbone ISPs with Netdiff. In: Proceedings of USENIX Symposium on Networked Systems Design and Implementation (NSDI) (2008)
72. R. Singh, A. Dunna, and P. Gill: Characterizing the Deployment and Performance of Multi-CDNs. In: Proceedings of ACM Internet Measurement Conference (IMC) (2018)
73. S. Biswas, J. Bicket, E. Wong, R. Musaloiu-E, A. Bhartia, and D. Aguayo: Large-scale Measurements of Wireless Network Behavior. In: Proceedings of ACM SIGCOMM (2015)
74. S. Grover, M. Park, S. Sundaresan, S. Burnett, H. Kim, and N. Feamster: Peeking Behind the NAT: An Empirical Study of Home Networks. In: Proceedings of ACM Internet Measurement Conference (IMC) (2013)
75. S. Guha, J. Chandrashekar, N. Taft, and K. Papagiannaki : How Healthy Are Today's Enterprise Networks? In: Proceedings of ACM Internet Measurement Conference (IMC) (2008)
76. S. Kandula, S. Sengupta, A. Greenberg, P. Patel, and R. Chaiken: The Nature of Data Center Traffic: Measurements & Analysis. In: Proceedings of ACM Internet Measurement Conference (IMC) (2009)
77. S. Narayana, A. Sivaraman, V. Nathan, P. Goyal, V. Arun, M. Alizadeh, V. Jeyakumar, and C. Kim: Language-Directed Hardware Design for Network Performance Monitoring. In: Proceedings of ACM SIGCOMM (2017)
78. S. Sundaresan, N. Feamster, R. Teixeira, and N. Magharei: Measuring and Mitigating Web Performance Bottlenecks in Broadband Access Networks. In: Proceedings of ACM Internet Measurement Conference (IMC) (2013)
79. S. Tao, K. Xu, A. Estepa, T. Fei, L. Gao, R. Guerin, J. Kurose, D. Towsley, and Z.-L. Zhang: Improving VoIP Quality Through Path Switching. In: Proceedings of IEEE INFOCOM (2005)
80. S. Tao, K. Xu, Y. Xu, T. Fei, L. Gao, R. Guerin, J. Kurose, D. Towsley, and Z.-L. Zhang: Exploring the Performance Benefits of End-to-End Path Switching. In: Proceedings of IEEE International Conference on Network Protocols (2004)
81. Shannon, C.E., Weaver, W.: The Mathematical Theory of Communication. University of Illinois Press (1949)
82. T. Høiland-Jørgensen, B. Ahlgren, P. Hurtig, and A. Brunstrom: Measuring Latency Variation in the Internet. In: Proceedings of ACM International Conference on emerging Networking EXperiments and Technologies (CoNEXT) (2016)
83. T. Qiu, J. Ni, H. Wang, N. Hua, Y. Yang, and J. Xu: Packet Doppler: Network Monitoring using Packet Shift Detection. In: Proceedings of ACM International Conference on emerging Networking EXperiments and Technologies (CoNEXT) (2008)
84. V. Demianiuk, S. Gorinsky, S. Nikolenko, and K. Kogan: Robust Distributed Monitoring of Traffic Flows. In: Proceedings of IEEE International Conference on Network Protocols (ICNP) (2019)
85. V. Sekar, M. Reiter, W. Willinger, H. Zhang, R. Kompella, and D. Andersen: CSAMP: A System for Network-Wide Flow Monitoring. In: Proceedings of USENIX Symposium on Networked Systems Design and Implementation (NSDI) (2008)
86. V. Vishnumurthy and P. Francis: On The Difficulty of Finding the Nearest Peer in P2P Systems. In: Proceedings of ACM Internet Measurement Conference (IMC) (2008)
87. Xu, K., Chandrashekar, J., Zhang, Z.L.: A First Step Towards Understanding Inter-domain Routing. In: Proceedings of ACM SIGCOMM Workshop on Mining Network Data (2005)
88. Y. Cheng, V. Ravindran, A. Leon-Garcia: Internet Traffic Characterization Using Packet-Pair Probing. In: Proceedings of IEEE INFOCOM (2007)
89. Y. Geng, S. Liu, Z. Yin, A. Naik, B. Prabhakar, M. Rosenblum, and A. Vahdat: SIMON: A Simple and Scalable Method for Sensing, Inference and Measurement in Data Center Networks. In: Proceedings of USENIX Symposium on Networked Systems Design and Implementation (NSDI) (2019)

90. Y. Huang, N. Feamster, A. Lakhina, and J. Xu: Diagnosing Network Disruptions with Network-Wide Analysis. In: Proceedings of ACM SIGMETRICS (2007)
91. Y. Shavitt and U. Weinsberg: Quantifying the Importance of Vantage Points Distribution in Internet Topology Measurements. In: Proceedings of IEEE INFOCOM (2009)
92. Y. Wang, L. Huang, T. Gu, H. Wei, K. Xing, and J. Zhang: Data-Driven Traffic Flow Analysis for Vehicular Communications. In: Proceedings of IEEE INFOCOM (2014)
93. Y. Zhao, Y. Chen, and D. Bindel: Towards Unbiased End-to-End Network Diagnosis. In: Proceedings of ACM SIGCOMM (2006)
94. Y. Zhou, Y. Zhang, C. Ma, S. Chen, and O. Odegbile: Generalized Sketch Families for Network Traffic Measurement. In: Proceedings of ACM SIGMETRICS (2020)
95. Z. Bischof, F. Bustamante, and R. Stanojevic: Need, Want, Can Afford – Broadband Markets and the Behavior of Users. In: Proceedings of ACM Internet Measurement Conference (IMC) (2014)
96. Z. Cai, M. Chen, S. Chen, and Y. Qiao: Searching for Widespread Events in Large Networked Systems by Cooperative Monitoring. In: Proceedings of IEEE International Conference on Network Protocols (ICNP) (2015)
97. Z. Duan, Z.-L. Zhang, Y. Hou, Service Overlay Networks: Assumptions and Bandwidth Provisioning Problems. IEEE/ACM Transactions on Networking **11**(6), 870–883 (2003)
98. Z.-L. Zhang, V. Ribeiro, S. Moon, and C. Diot: Small-Time Scaling Behaviors of Internet Backbone Traffic: An Empirical Study. In: Proceedings of IEEE INFOCOM (2003)
99. Z. Liu, A. Manousis, G. Vorsanger, V. Sekar, and V. Braverman: One Sketch to Rule Them All: Rethinking Network Flow Monitoring with UnivMon. In: Proceedings of ACM SIGCOMM (2016)

Chapter 3
Behavior Modeling of Network Traffic

Abstract Given the size and scale of the Internet, networked systems and Internet applications often send and receive a large number of data packets with a wide spectrum of traffic features, such as unique numbers of source and destination IP addresses, unique numbers of source and destination ports, and total numbers of flows, packets, and bytes. Clearly, it is impractical for network operators and security analysts to examine every IP data packet or network flow in order to understand communication patterns and traffic behaviors of networked systems and Internet applications. Thus, we need effective models, techniques, and tools to analyze and summarize traffic behaviors on the Internet. This chapter starts with an introduction of the concepts, features, and models of network behavior, and then introduces an entropy-based adaptive thresholding algorithm to extract significant networked systems and Internet applications. Subsequently, this chapter presents an *information-theoretic* approach to characterize traffic patterns of behavioral entities on the Internet and shows that this leads to a natural behavioral classification scheme for grouping networked systems and Internet applications into *behavior classes* (BC) with distinct behavior patterns. By examining the characteristics of these behavior classes and individual hosts and applications over time, this chapter also sheds light on the stability and temporal dynamics of network behavior and traffic patterns of networked systems and Internet applications.

3.1 Behavior-Oriented Network Traffic Modeling

3.1.1 What is Network Behavior

As the Internet continues to grow in users, devices, applications, Services, and traffic, *what is happening on the internet* has become a daunting task for network operators and security analysts. The concept of *network behavior* refers to the behaviors of networked systems and Internet applications, which are observable from and embedded

K. Xu, *Network Behavior Analysis*,
https://doi.org/10.1007/978-981-16-8325-1_3

in the underlying traffic data exchanged between networked systems on the Internet. Understanding network behavior from unstructured traffic data can not only tell who talks to whom on the Internet, when does the conversation happen, for how long, and why, but also answer many basic and advanced questions such as what are the top ten Internet applications measured by network traffic volume and where the most aggressive cyber attack traffic comes from. Given the wide spectrum of critical applications in network operations and engineering, *network behavior analysis* has become a critical research field in the networking, system, and security research communities.

3.1.2 Traffic Features in Network Behavior

Networked systems communicate with each other via exchanging TCP/IP data packets on the Internet. Each data packet consists of a network layer header and packet payload which encapsulates data from the transport layer such as TCP and UDP segments [1]. A single data communication or "conversation" between two networked systems on the Internet could incur hundreds or thousands of IP data packets that share many fields in the IP, TCP, or UDP headers. Therefore, aggregating IP data packets from the same *conversation* into a network *flow* is a common practice in network traffic analysis. All the operating systems of major internet routers such as Cisco IOS Software [2] and Junos OS [3] embed and support real-time instrumentation of exporting network flows for many mission-critical applications such as network engineering, capacity planning, performance monitoring, network troubleshooting, anomaly detection, and network behavior analysis.

A network flow is defined based on the well-known 5-tuple, i.e., the source IP address (`srcIP`), destination IP address (`dstIP`), source port number (`srcPort`), destination port number (`dstPort`), and protocol, which collectively describe a unique *conversation* between two networked systems [4]. In addition to these five key traffic features characterizing network flows, the start and end timestamps of the flow, packet and byte counts of the flow often provide valuable information and critical insights into the conversation between networked systems. In the literature of network behavior analysis, the three terms—traffic features, traffic dimensions, and traffic fields are often used exchangeably.

3.1.3 Behavioral Entities

Similar to the classic 5 W questions, i.e., who, what, where, when, and why, in story writing and telling, network behavior analysis characterizes and interprets *who*, *what*, *when*, *where*, and *why* of data communications between networked systems on the Internet. The *who* element in network behavior analysis represents the source and the destination of data communications, and *what* represents the Internet applications and services of the communication. The *when* element captures the start and end timestamps of data communications, while *where* means the geographical locations of the source and destinations. The final *why* element often requires careful investigations, analysis, and mining on the current and historical network traffic for establishing plausible interpretations of each data communication between networked systems.

In general, the behavioral entities, also referred to as behavioral objects and behavioral units, in network traffic refer to networked systems and Internet applications. The networked systems observed in TCP/IP packets or network flows are `srcIP` and `dstIP`, while the Internet applications in IP packets or network flows are `srcPort` and `dstPort`. To extract network traffic for individual behavioral entities, we start by slicing network traffic along each dimension of the four-feature space, `srcIP, dstIP, srcPrt`, or `dstPrt`, and build traffic clusters for these four dimensions. Making sense of the extracted `srcIP` and `dstIP` clusters yields the host behaviors and communication patterns for a set of networked systems, while understanding the `srcPort` and `dstPort` clusters yields the port behaviors for Internet applications and services which reflect the aggregate behaviors of individual end hosts on the corresponding application ports. Next, we present a multidimensional view of network traffic data for modeling network behaviors of these entities.

3.1.4 Real-World Network Traffic Datasets

To systematically evaluate network behavior modeling, we obtain real-world network traffic datasets from multiple links in a large ISP network at the core of the Internet (Table 3.1). For every 5-min time slot, we aggregate packet header (the first 40 bytes of each packet) traces into network flows with a timeout value of 60 s [5]. The 5-min time slot is used as a trade-off between timeliness of traffic behavior profiling and the amount of data to be processed in each slot.

As shown in Table 3.1, the packet trace lengths range in duration from 3 to 24 h, and the links vary in capacity from 155 Mbps to 10 Gbps. Moreover, they carry diverse types of traffic—*access* links carry traffic to and from a set of customers, *wireless* links carry traffic between a 3G wireless data network and the rest of the Internet, while *backbone* links carry a mix of customer and peer traffic in the middle of the ISP's network. These links have been chosen carefully in order to demonstrate the general applicability of our approach to a number of representative traffic mixes.

Table 3.1 Multiple links used in network behavior analysis

Link	Link type	Utilization	Duration	Packets	Trace size
L_1	Backbone	78 Mbps	24 h	$1.60 * 10^9$	95 GB
L_2	Access	86 Mbps	24 h	$1.65 * 10^9$	98 GB
L_3	Access	40 Mbps	3 h	$2.03 * 10^8$	12 GB
L_4	Access	52 Mbps	3 h	$1.91 * 10^8$	11 GB
L_5	Wireless	207 Mbps	3 h	$5.18 * 10^8$	28 GB

3.2 Identifying Significant Behavioral Entities

3.2.1 Significant Behavioral Entities

Network traffic on the Internet typically have a very diverse traffic mixes since the Internet carries network traffic for different networked systems on a large number of network applications in a very short time window. Thus, it is not practical to analyze the mixed behavior of all observed networked systems and network applications. As a result, network behavioral analysis often focuses on analyzing network traffic of significant behavioral entities, i.e., individual networked systems and network applications, which are associated with a significant amount of traffic.

Using the four-dimensional (`srcIP, dstIP, srcPort, and dstPort`) feature space, we extract "clusters" of significance along each dimension, where each cluster consists of flows with the same feature value in the said dimension. This results in four collections of significant traffic clusters—`srcIP` clusters, `dstIP` clusters, `srcPrt` clusters, and `dstPrt` clusters. The `srcIP` and `dstIP` clusters represent the collections of end hosts, while the `srcPort` and dstPort clusters represent the collections of network applications or services.

Traditional approaches in extracting traffic clusters of significance rely on a fixed threshold based on volume, such as packet, byte, and flow counts [4]. However, given the diverse traffic mixes in hundreds of Internet links in an IP backbone network, simple fixed thresholds could not fit well into all the links. Therefore, we develop an information-theoretic approach that culls interesting clusters based on the underlying feature value distribution (or *entropy*) in the fixed dimension. Intuitively, clusters with feature values (cluster keys) that are distinct in terms of distribution are considered significant and extracted; this process is repeated until the remaining clusters appear indistinguishable from each other. This yields a cluster extraction algorithm that automatically adapts to the traffic mix and the feature in consideration.

3.2.2 *Adaptive Thresholding Algorithm*

We start by focusing on each dimension of the four-feature space, `srcIP`, `dstIP`, `srcPrt`, or `dstPrt`, and extract significant clusters of interest along this dimension. The extracted `srcIP` and `dstIP` clusters yield a set of interesting host behaviors (communication patterns), while the `srcPrt` and `dstPrt` clusters yield a set of interesting service/port behaviors, reflecting the aggregate behaviors of individual hosts on the corresponding ports. In the following, we introduce our definition of *significance/interestingness* using the (conditional) relative uncertainty measure (cf. Appendix A).

Given one feature dimension X and a time interval T, let m be the total number of flows observed during the time interval, and $A = \{a_1, \ldots, a_n\}$, $n \geq 2$, be the set of distinct values (e.g., `srcIP`'s) in X that the observed flows take. Then the (induced) probability distribution $\mathcal{P}_A$ on X is given by $p_i := \mathcal{P}_A(a_i) = m_i/m$, where m_i is the number of flows that take the value a_i (e.g., having the `srcIP` a_i). Then the (conditional) relative uncertainty, $RU(\mathcal{P}_A) := RU(X|A)$, measures the degree of uniformity in the observed features A. If $RU(\mathcal{P}_A)$ is close to 1, say, $> \beta = 0.9$, then the observed values are close to being uniformly distributed, and thus *nearly indistinguishable*. Otherwise, there are likely feature values in A that "stand out" from the rest. We say a subset S of A contains the *most significant* (thus "interesting") values of A if S is the smallest subset of A such that (i) the probability of any value in S is larger than those of the remaining values; and (ii) the (conditional) probability distribution on the set of the remaining values, $R := A - S$, is close to being uniformly distributed, i.e., $RU(\mathcal{P}_R) := RU(X|R) > \beta$. Intuitively, S contains the most significant feature values in A, while the remaining values are nearly indistinguishable from each other.

To see what S contains, order the feature values of A based on their probabilities: let $\hat{a}_1, \hat{a}_2, \ldots, \hat{a}_n$ be such as $\mathcal{P}_A(\hat{a}_1) \geq \mathcal{P}_A(\hat{a}_2) \geq \cdots \mathcal{P}_A(\hat{a}_n)$. Then $S = \{\hat{a}_1, \hat{a}_2, \ldots, \hat{a}_{k-1}\}$, and $R = A - S = \{\hat{a}_k, \hat{a}_{k+1}, \ldots, \hat{a}_n\}$, where k is the smallest integer such that $RU(\mathcal{P}_R) > \beta$. Let $\alpha^* = \hat{a}_{k-1}$. Then α^* is the largest "cut-off" threshold such that the (conditional) probability distribution on the set of remaining values R is close to being uniformly distributed. To extract S from A (thereby, the clusters of flows associated with the significant feature values), we take advantage of the fact that in practice the probability distribution of the feature values $\mathcal{P}_A$ in general obeys a *power-law*: only a relatively few values (with respect to n) have significant larger probabilities, i.e., $|S|$ is relatively small, while the remaining feature values are close to being uniformly distributed. Hence, we can efficiently search for the optimal cut-off threshold α^*.

Algorithm 1 presents an efficient *approximation* algorithm (in pseudo-code) for extracting the significant clusters in S from A (thereby, the clusters of flows associated with the significant feature values). The algorithm starts with an appropriate initial value α_0 (e.g., $\alpha_0 = 2\%$), and searches for the optimal cut-off threshold α^* from above via "exponential approximation" (reducing the threshold α by an exponentially decreasing factor $1/2^k$ at the kth step). As long as the relative uncertainty of the

Algorithm 3.1 Entropy-based Significant Cluster Extraction

1: Parameters: $\alpha := \alpha_0$; $\beta := 0.9$; $S := \emptyset$;
2: Initialization: $S := \emptyset$; $R := A$; $k := 0$;
3: compute prob. dist. $\mathcal{P}_R$ and its RU $\theta := RU(\mathcal{P}_R)$;
4: **while** $\theta \leq \beta$ **do**
5: $\alpha = \alpha \times 2^{-k}$; $k++$;
6: **for each** $a_i \in R$ **do**
7: **if** $\mathcal{P}_A(a_i) \geq \alpha$ **then**
8: $S := S \cup \{a_i\}$; $R := R - \{a_i\}$;
9: **end if**
10: **end for**
11: compute (cond.) prob. dist. $\mathcal{P}_R$ and $\theta := RU(\mathcal{P}_R)$;
12: **end while**

(conditional) probability distribution $\mathcal{P}_R$ on the (remaining) feature set R is less than β, the algorithm examines each feature value in R and includes those whose probabilities exceed the threshold α into the set S of significant feature values. The algorithm stops when the probability distribution of the remaining feature values is close to being uniformly distributed ($>\beta = 0.9$). Let $\hat{\alpha}^*$ be the final cut-off threshold (an approximation to α^*) obtained by the algorithm.

3.2.3 Extracting Significant Traffic Clusters

Our algorithm adaptively adjusts the "cut-off" threshold $\hat{\alpha}^*$ based on the underlying feature value distributions to extract significant clusters. In our study, we have found that $\alpha_0 = 0.02$ provides a good starting point, and $\hat{\alpha}^*$ is typically in the range of $[0.0001, 0.02]$ based on the feature value distributions.

Figure 3.1 presents the results we obtain by applying the algorithm to the 24-h packet trace collected on L_1, where the significant clusters are extracted in every 5-min time slot along each of the four feature dimensions. In each plot, we show both the total number of distinct feature values as well as the number of significant clusters extracted in each 5-min slot over 24 h for the four feature dimensions (note that the y-axis is in log scale). We see that, while the total number of distinct values along a given dimension may not fluctuate very much, the number of significant feature values (clusters), which is less than the total number of distinct values in several orders of magnitude, may vary dramatically, due to changes in the underlying feature value distributions.

Figure 3.2 shows the corresponding final cut-off threshold obtained by the algorithm. It is very interesting to observe that different cut-off thresholds being used in extracting the significant feature values (clusters) during the 24-h period, which suggests that the entropy-based thresholding algorithm could adjust to the underlying traffic patterns. In fact, the dramatic changes in the number of significant clusters (or equivalently, the cut-off threshold) also signifies major changes in the underlying

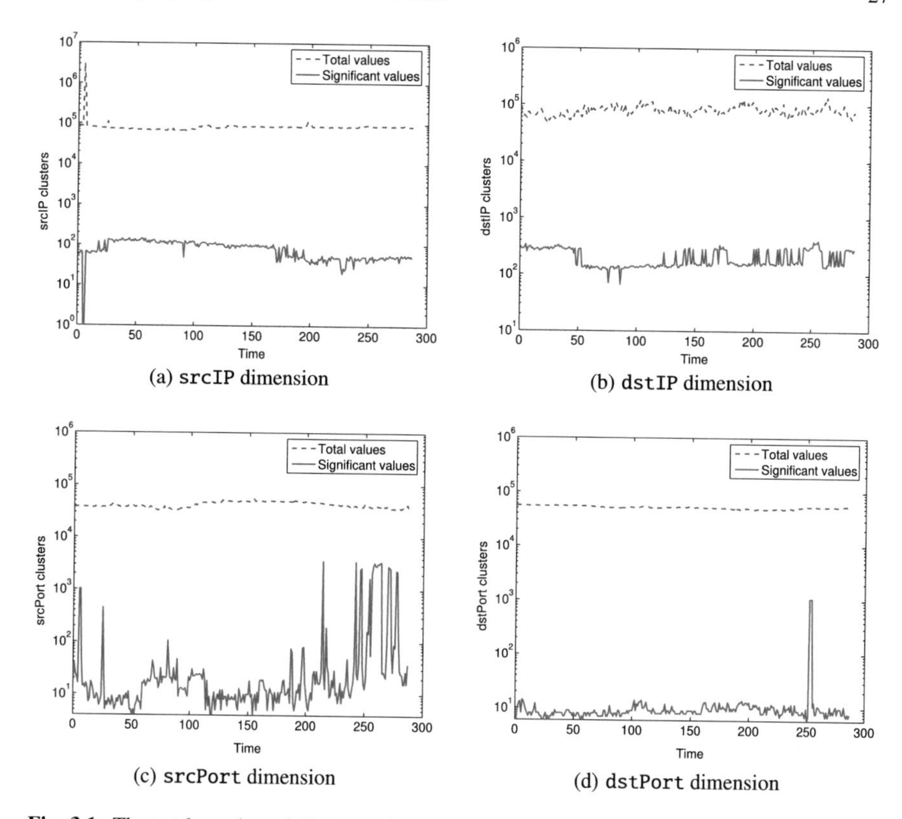

(a) `srcIP` dimension
(b) `dstIP` dimension
(c) `srcPort` dimension
(d) `dstPort` dimension

Fig. 3.1 The total number of distinct values and significant clusters extracted from four feature dimensions of L_1 over a 1-day period

Table 3.2 The number of significant clusters extracted from L_1 during the 15th time period using adaptive threshold

Dimension	# Distinct values	Threshold	# significant clusters
`srcIP`	89261	0.0625%	117
`dstIP`	79660	0.03125%	273
`srcPort`	49511	0.25%	8
`dstPort`	50602	1%	12

traffic patterns. In addition, the adaptive thresholds across four feature dimensions are also different, and indicate that fixed thresholds for multiple dimensions will often fail to uncover different size distributions for four traffic dimensions.

To provide some specific numbers, consider the 15th time slot. As shown in Table 3.2, there are a total of 89261 distinct `srcIP`'s, 79660 distinct `dstIP`'s, 49511 `srcPrt`'s, and 50602 `dstPrt`'s. Our adaptive-threshold algorithm extracts 117 significant `srcIP` clusters, 273 `dstIP` clusters, 8 `srcPrt` clusters and 12

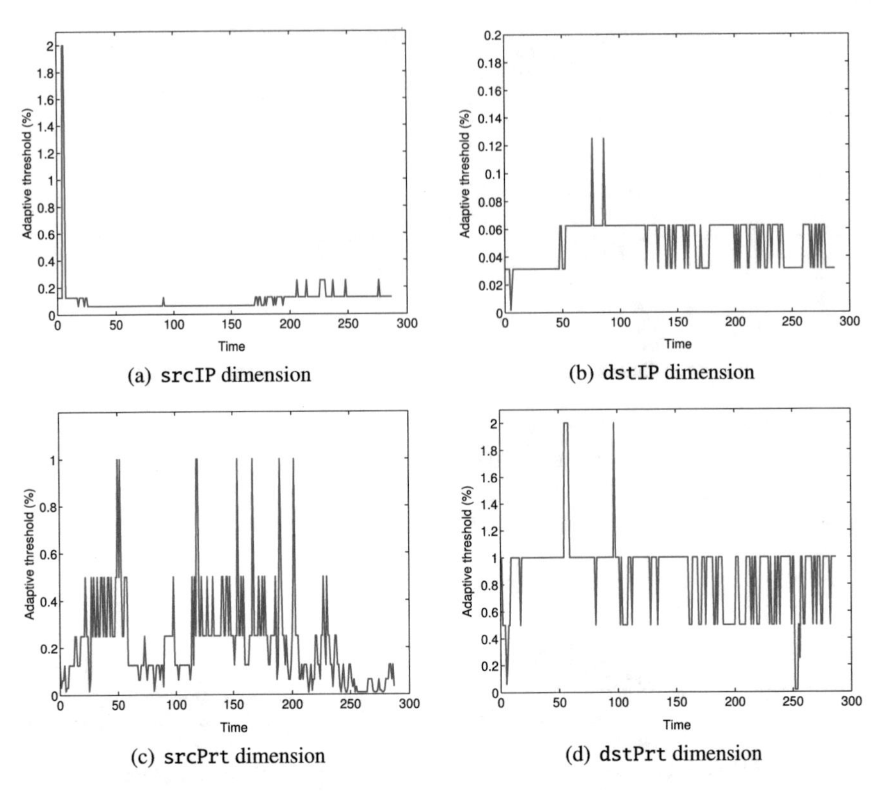

(a) `srcIP` dimension (b) `dstIP` dimension (c) `srcPrt` dimension (d) `dstPrt` dimension

Fig. 3.2 The final cut-off threshold of four feature dimensions of L_1 over a 1-day period obtained by the cluster extraction algorithm

`dstPrt` clusters, with the resulting cut-off threshold being 0.0625%, 0.03125%, 0.25%, and 1%, respectively. We see that the number of significant clusters is far smaller than the number of feature values, and that the cut-off thresholds for the different feature dimensions also differ. This shows that *no* single *fixed* threshold would be adequate in the definition of "significant" behavior clusters. In fact, we will show later that significant clusters themselves are quite diverse in their sizes, whether measured in the number of flows (flow count), packets (packet count), or bytes (byte count). Therefore, focusing only on top clusters based purely on volumes may miss many otherwise significant behaviors that are interesting, rare or anomalous, and thus warrant special attention.

Table 3.3 Convention of free dimension denotations in traffic clusters

Cluster key	Free dimensions		
	X	Y	Z
srcIP	srcPrt	dstPrt	dstIP
dstIP	srcPrt	dstPrt	srcIP
srcPrt	dstPrt	srcIP	dstIP
dstPrt	srcPrt	srcIP	dstIP

3.3 Network Behavior Modeling

3.3.1 *Network Behavior Modeling*

Modeling network behavior for the clusters of networked systems and Internet applications start from characterizing traffic features of data communications in each cluster and capturing the interactions among these features. Consider the set of, say, `srcIP`, clusters extracted from flows observed in a given time slot. The flows in each `srcIP` cluster share the same behavioral entity, also called cluster key, while they can take any possible value along the other three free dimensions, i.e., four basic dimensions except the cluster dimension. In this case, `dstIP`, `srcPort`, and `dstPort` are free dimensions. Hence, the flows in a cluster induce a probability distribution on each of the three "free" dimensions, and thus a *relative uncertainty* (cf. Chap. 2) measure can be defined.

For each cluster extracted along a fixed dimension, we use X, Y and Z to denote its three "free" dimensions, using the convention listed in Table 3.3. Hence, for a `srcIP` cluster, X, Y, and Z denote the `srcPrt`, `dstPrt`, and `dstIP` dimensions, respectively. This cluster can be characterized by an *RU vector* $[RU_X, RU_Y, RU_Z]$. In other words, the multidimensional RU vector characterizes the traffic patterns of the behavioral entity, i.e., the cluster key.

In Fig. 3.3a, we represent the RU vector of each `srcIP` cluster extracted in each 5-min time slot over a 1-h period from L_1 as a point in a unit-length cube. We see that most points are "clustered" (in particular, along the axes), suggesting that there are certain common "behavior patterns" among them. Fig. 3.4 shows similar results using the `srcIP` clusters on four other links. This "clustering" effect can be explained by the "multi-modal" distribution of the relative uncertainty metrics along each of the three free dimensions of the clusters, as shown in Fig. 3.3b–d where we plot the histogram (with a bin size of 0.1) of RU_X, RU_Y, and RU_Z of all the clusters on links L_1–L_5, respectively. For each free dimension, the RU distribution of the clusters is multi-modal, with two strong modes (in particular, in the case of `srcPrt` and `dstPrt`) residing near the two ends, 0 and 1. Similar observations also hold for `dstIP`, `srcPrt`, and `dstPrt` clusters extracted on these links.

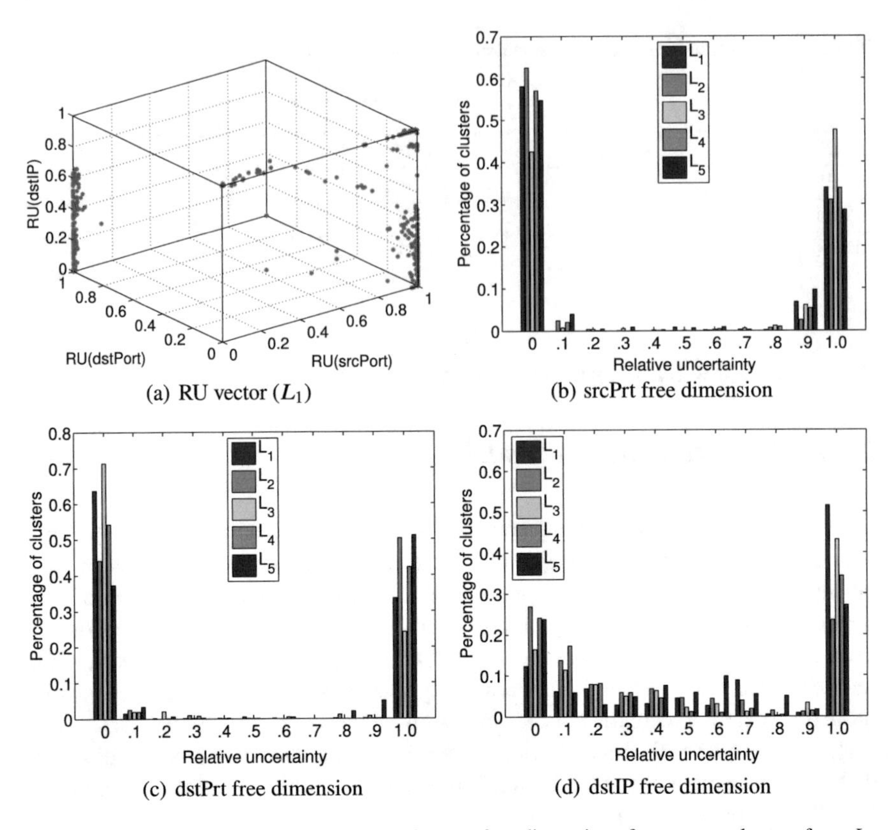

(a) RU vector (L_1)
(b) srcPrt free dimension
(c) dstPrt free dimension
(d) dstIP free dimension

Fig. 3.3 The distribution of relative uncertainty on free dimensions for `srcIP` clusters from L_1 during a 1-h period

3.3.2 Network Behavior Classifications

As a *convenient* way to group together clusters of similar behaviors, we divide each RU dimension into three categories (assigned with a label): 0 (low), 1 (medium), and 2 (high), using the following criteria:

$$L(ru) = \begin{cases} 0(low), & \text{if } 0 \leq ru \leq \epsilon, \\ 1(medium), & \text{if } \epsilon < ru < 1 - \epsilon, \\ 2(high), & \text{if } 1 - \epsilon \leq ru \leq 1, \end{cases} \tag{3.1}$$

where for the `srcPrt` and `dstPrt` dimensions, we choose $\epsilon = 0.2$, while for the `srcIP` and `dstIP` dimensions, $\epsilon = 0.3$. This labeling process classifies clusters into 27 possible *behavior classes* (*BC* in short), each represented by a (label) vector $[L(RU_X), L(RU_Y), L(RU_Z)] \in \{0, 1, 2\}^3$. For ease of reference, we also treat $[L(RU_X), L(RU_Y), L(RU_Z)]$ as an integer (in ternary representation)

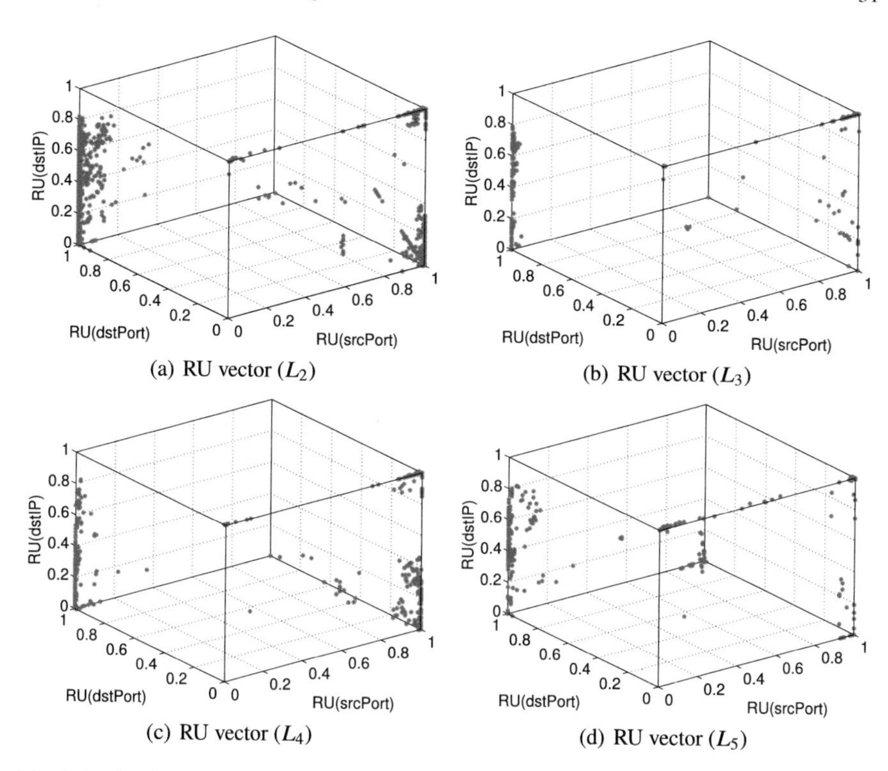

(a) RU vector (L_2)
(b) RU vector (L_3)
(c) RU vector (L_4)
(d) RU vector (L_5)

Fig. 3.4 The distribution of relative uncertainty on free dimensions for srcIP clusters from $L_{2,3,4,5}$ during a 1-h period

$id = L(RU_X) \cdot 3^2 + L(RU_Y) \cdot 3 + L(RU_Z) \in \{0, 1, 2, \ldots, 26\}$, and refer to it as BC_{id}. Hence `srcIP` $BC_6 = [0, 2, 0]$, which intuitively characterizes the communicating behavior of a networked system using a single or a few `srcPrt`'s to talk with a single or a few `dstIP`'s on a larger number of `dstPrt`'s. We remark here that for clusters extracted using other *fixed* feature dimensions (e.g., `srcPrt`, `dstPrt` or `dstIP`), the BC labels and id's have a different meaning and interpretation, as the free dimensions are different (see Table 3.3). We will explicitly refer to the BCs defined along each dimension as `srcIP` BCs, `dstIP` BCs, `srcPrt` BCs and `dstPrt` BCs. However, when there is no confusion, we will drop the prefix.

3.4 Network Behavior Dynamics

3.4.1 Temporal Properties of Behavior Classes

The RU-based behavioral analysis leads to a natural classification of network behaviors and communication patterns of networked systems and Internet applications. However, such behavioral analysis relies on traffic features from datasets collected at a certain time window. Thus, it is crucial to analyze the temporal properties of the behavior classes for the time-series analysis, stability analysis and change detection of network behavior.

Towards this end, we introduce three metrics to capture three different aspects of the characteristics of the BC's over time: (i) *popularity*: which is the number of times we observe a particular BC appearing (i.e., at least one cluster belonging to the BC is observed); (ii) *(average) size*: which is the average number of clusters belonging to a given BC, whenever it is observed; and (iii) *(membership) volatility*: which measures whether a given BC tends to contain the same clusters over time (i.e., the member clusters re-appear over time), or new clusters.

Formally, consider an observation period of T time slots. For each BC_i, let C_{ij} be the number of observed clusters that belong to BC_i in the time slot τ_j, $j = 1, 2, \ldots, T$, O_i the number of time slots that BC_i is observed, i.e.,

$$O_i = |\{C_{ij} : C_{ij} > 0\}|, \tag{3.2}$$

and U_i be the number of *unique clusters* belonging to BC_i over the entire observation period. Then the popularity of BC_i is defined as:

$$\Pi_i = \frac{O_i}{T}; \tag{3.3}$$

its average size Σ_i is defined as:

$$\Sigma_i = \sum_{j=1}^{T} \frac{C_{ij}}{O_i}; \tag{3.4}$$

and its (membership) volatility Ψ_i is defined as:

$$\Psi_i = \frac{U_i}{\sum_{j=1}^{T} C_{ij}} = \frac{U_i}{\Sigma_i O_i}. \tag{3.5}$$

If a BC contains the same clusters in all time slots, i.e., $U_i = C_{ij}$, for every j such that $C_{ij} > 0$, then $\Psi_i = 0$. In general, the closer Ψ_i is to 0, the less volatile the BC is. Note that the membership volatility metric is defined only for BC's with relatively high frequency, e.g., $\Pi > 0.2$, as otherwise it contains too few "samples" to be meaningful.

In Fig. 3.5a–c we plot Π_i, Σ_i and Ψ_i of the `srcIP` BC's for the `srcIP` clusters extracted using link L_1 over a 24-h period, where each time slot is a 5-min interval (i.e., $T = 288$). From Fig. 3.5a we see that 7 BC's, BC_2 [0, 0, 2], BC_6 [0, 2, 0], BC_7 [0, 2, 1], BC_8 [0, 2, 2], BC_{18} [2, 0, 0], BC_{19} [2, 0, 1] and BC_{20} [2, 0, 2], are most popular, occurring more than half of the time; while BC_{11} [2, 0, 2] and BC_{12} [2, 1, 0] and BC_{24} [2, 2, 1] have moderate popularity, occurring about one-third of the time. The remaining BC's are either rare or not observed at all. Figure 3.5b shows that the five popular BC's, BC_2, BC_6, BC_7, BC_{18}, and BC_{20}, have the largest (average) size, each having around 10 or more clusters; while the other two popular BC's, BC_8 and BC_{19}, have four or fewer BC's on the average. The less popular BC's are all small, having at most one or two clusters on the average when they are observed. From Fig. 3.5c, we see that the two popular BC_2 and BC_{20} (and the less popular BC_{11}, BC_{12} and BC_{24}) are most volatile, while the other five popular BC's, BC_6, BC_7, BC_8, BC_{18} and BC_{19} are much less volatile. To better illustrate the difference in the membership volatility of the 7 popular BC's, in Fig. 3.5d we plot U_i as a function of time, i.e., $U_i(t)$ is the total number of unique clusters belonging to BC_i *up to time slot* t. We see that for BC_2 and BC_{20}, new clusters show up in nearly every time slot, while for BC_7, BC_8 and BC_{19}, the same clusters re-appear again and again. For BC_6 and BC_{18}, new clusters show up gradually over time and they tend to re-occur, as evidenced by the tapering off of the curves and the large average size of these two BC's.

3.4.2 Behavior Dynamics of Individual Clusters

We now investigate the behavior characteristics of individual clusters over time. In particular, we are interested in understanding (i) the relation between the *frequency* of a cluster (i.e., how often it is observed) and the behavior class(es) it appears in; and (ii) the behavior *stability* of a cluster if it appears multiple times, namely, whether a cluster tends to re-appear in the same BC or different BC's?

We use the set of `srcIP` clusters extracted on links with the longest duration, L_1 and L_2, over a 24-h period as two representative examples to illustrate our findings. Figure 3.6 shows the frequency distribution of clusters in *log-log* scale, where the x-axis is the cluster id ordered based on its frequency (the most frequent cluster first). The distribution is "heavy-tailed": for example more than 90.3% (and 89.6%) clusters in L_1 (and L_2) occur fewer than 10 times, of which 47.1% (and 55.5%) occur only once; 0.6% (and 1.2%) occur more than 100 times. Moreover, the most frequent clusters all fall into the five popular but non-volatile BC's, BC_6, BC_7, BC_8, BC_{18} and BC_{19}, while a predominant majority of the least frequent clusters belong to BC_2 and BC_{20}. The medium-frequency clusters belong to a variety of BCs, with BC_2 and BC_{20} again dominant.

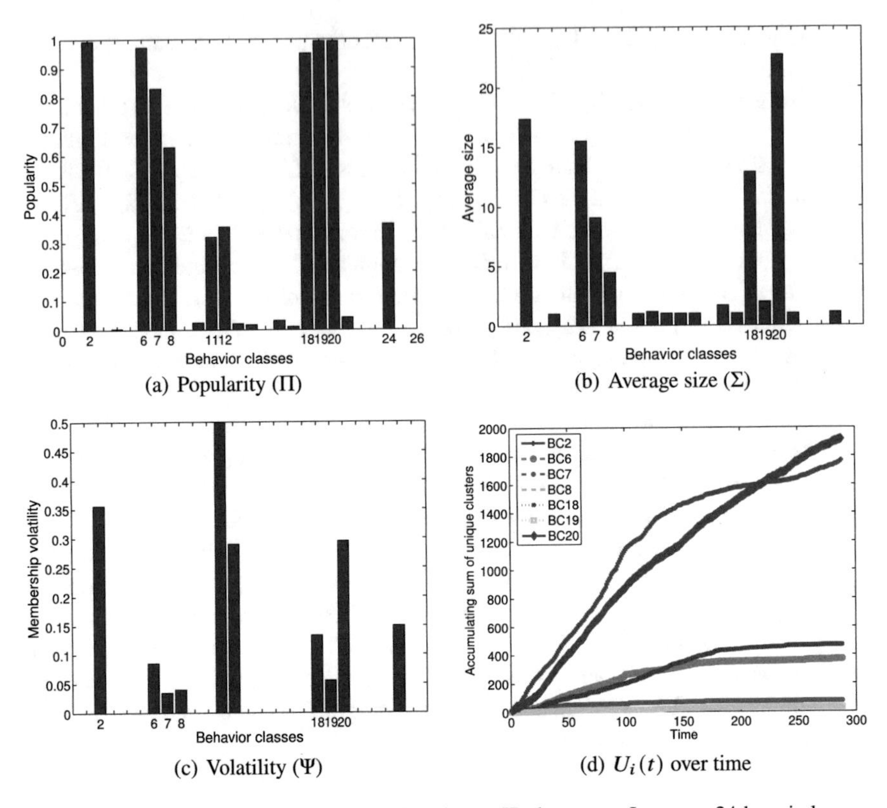

Fig. 3.5 Temporal properties of srcIP BCs using srcIP clusters on L_1 over a 24-h period

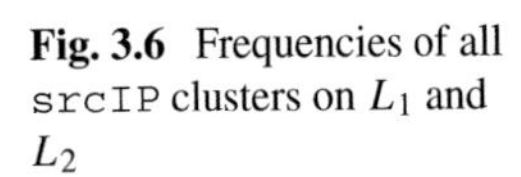

Fig. 3.6 Frequencies of all srcIP clusters on L_1 and L_2

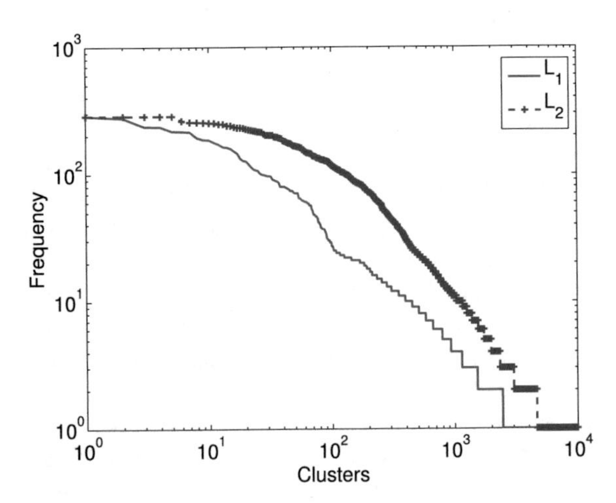

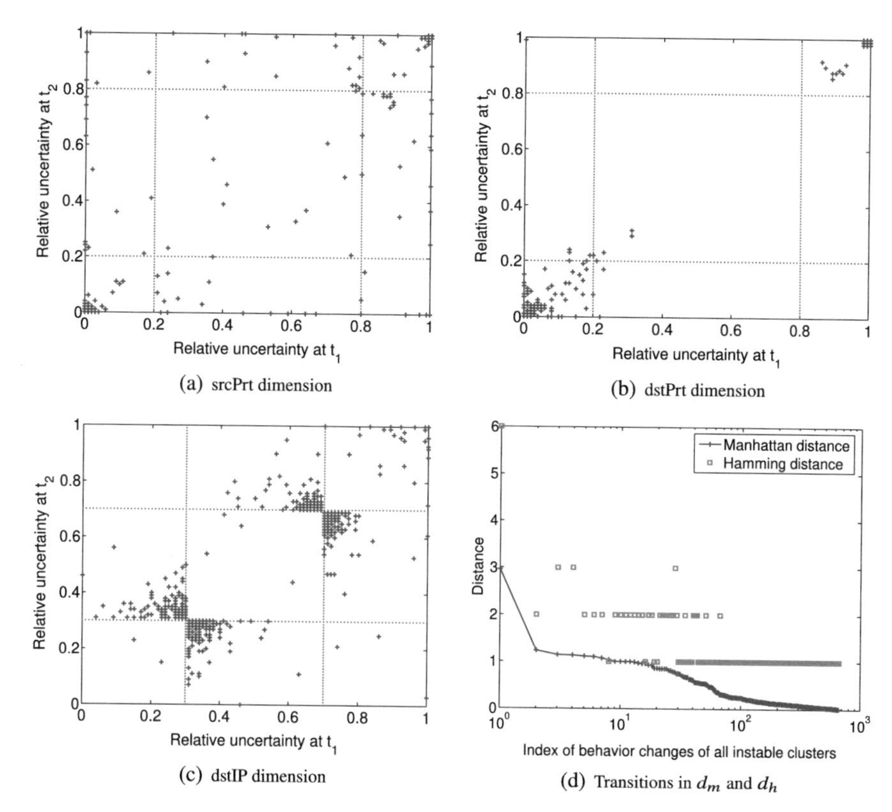

Fig. 3.7 Behavior transitions along `srcPrt`, `dstPrt` and `dstIP` dimensions as well as Manhattan and Hamming distances for "multi-BC" `srcIP` clusters on L_1

Next, for those clusters that appear at least twice (2,443 and 4,639 `srcIP` clusters from link L_1 and L_2, respectively), we investigate whether they tend to re-appear in the same BC or different BC's. We find that a predominant majority (nearly 95% on L_1 and 96% on L_2) stay in the same BC when they re-appear. Only a few (117 clusters on L_1 and 337 on L_2) appear in more than 1 BC. For instance, out of the 117 clusters on L_1, 104 appear in 2 BC's, 11 in 3 BC's and 1 in 5 BC's. We refer to these clusters as "multi-BC" clusters.

We perform an in-depth analysis on the "behavior transitions" of these "multi-BC" clusters in terms of their RU vectors (RUVs). In Fig. 3.7 we examine the behavior transitions of those 117 "multi-BC" clusters along each of the three dimensions (`srcPrt`, `dstPrt` and `dstIP`), where each point represents an RU transition $(RU(t_1), RU(t_2))$ in the corresponding dimension. We see that for each dimension, most of the points center around the diagonal, indicating that the RU values typically do not change significantly. For those transitions that cross the boundaries, causing a BC change for the corresponding cluster, most fall into the rectangle boxes along the sides, with only a few falling into the two square boxes on the upper left and lower right corners. This means that along each dimension, most of the BC changes can be attributed to transitions between two adjacent labels.

To measure the combined effect of the three RU dimensions on behavior transitions, we define two distance metrics: *Manhattan distance* (d_m) and *Hamming distance* (d_h):

$$\begin{aligned} d_m = &|RU_X(t_1) - RU_X(t_2)| + |RU_Y(t_1) - RU_Y(t_2)| \\ &+ |RU_Z(t_1) - RU_Z(t_2)|, \end{aligned} \tag{3.6}$$

and

$$\begin{aligned} d_h = &|L_X(t_1) - L_X(t_2)| + |L_Y(t_1) - L_Y(t_2)| \\ &+ |L_Y(t_1) - L_Y(t_2)|, \end{aligned} \tag{3.7}$$

where L is the labeling function (cf., Eq. (3.1)).

Figure 3.7d plots the Manhattan distance and Hamming distance of those behavior transitions that cause a BC change (a total of 658 such instances) for one of the "multi-BC" clusters. These behavior transitions are indexed in the decreasing order of Manhattan distance. We see that over 90% of the "BC-changing" behavior transitions have only a small Manhattan distance (e.g., $\leq$0.4), and most of the BC changes are within *akin* BC's, i.e., with a Hamming distance of 1. Only 60 transitions have a Manhattan distance larger than 0.4, and 31 have a Hamming distance of 2 or 3, causing BC changes between *non-akin* BC's. Hence, in a sense, only these behavior transitions reflect a large deviation from the norm. These "deviant" behavior transitions can be attributed to large RU changes in the `srcPrt` dimension, followed by the `dstIP` dimension. Out of the 117 "multi-BC clusters, we find that only 28 exhibit one or more "deviant" behavior transitions (i.e., with $d_m \geq 0.4$ or $d_h = 2, 3$) due to significant traffic pattern changes, and thus are regarded as *unstable* clusters. The above analysis has therefore enabled us to distinguish between this small set of clusters from the rest of the multi-BC clusters for which behavior transitions are between *akin* BCs, and a consequence of the choice of ϵ in Eq. (3.1), rather than any significant behavioral changes.

We conclude the analysis on network behavior dynamics by commenting that our observations and results regarding the temporal properties of behavior classes and behavior dynamics of individual clusters hold not only for the `srcIP` clusters extracted on L_1 but also on other dimensions and links we studied. Furthermore, qualitatively similar observations can also be made regarding the behavior characteristics of `dstIP`, `srcPrt` and `dstPrt` BCs and individual clusters.

3.5 Summary

This chapter demonstrates that behavior modeling of Internet traffic characterizes and differentiates behavioral patterns of networked systems and Internet applications via capturing the probability distributions and interactions of a wide spectrum of traffic features in TCP/IP data packets and network flows. The behavior classes defined by our entropy-based behavior classification scheme manifest distinct temporal characteristics, as captured by the frequency, populousness, and volatility metrics. In addition, traffic clusters formed by behavioral entities such as networked systems and Internet applications in general exhibit consistent behaviors over time, with only a very few occasionally exhibiting unstable behaviors.

In summary, the RU-based behavior classification scheme inherently captures certain behavior similarity among traffic clusters. This similarity is in essence measured by how varied (e.g., random or deterministic) the flows in a cluster assume feature values in the other three free dimensions. Thus, the resulting behavior classification provides meaningful and quantifiable measures for characterizing individual behaviors of networked systems and Internet applications during both the short-time snapshot and the long-term observations of network traffic. In other words, the entropy-based behavior classification scheme is a robust and consistent behavior model of network traffic. The next two chapters will present how to develop structural and graphical models, complement to RU-based behavioral model, to make sense of network data traffic for networked systems and Internet applications.

References

1. J. Postel, RFC791: Internet Protocol (1981)
2. C. Systems, Cisco ios software, https://www.cisco.com/c/en/us/products/ios-nx-os-software/ios-software-releases-listing.html
3. J. Networks, Junos os, https://www.juniper.net/documentation/product/en_US/junos-os
4. C. Estan, S. Savage, G. Varghese, Automatically inferring patterns of resource consumption in network traffic, in *Proceedings of ACM SIGCOMM* (2003)
5. K. Claffy, H.-W. Braun, G. Polyzos, A parameterizable methodology for internet traffic flow profiling. IEEE J. Sel. Areas Commun. (1995)

Chapter 4
Structural Modeling of Network Traffic

Abstract The behavior modeling of network traffic characterizes the distributions of traffic features for networked systems and Internet applications but sheds little light on the interactions between traffic features. The structural modeling of network traffic fills this void by capturing the interactions between features and identifying the dominant feature values, if they exist, for describing the communication structure and activity for the behavioral entities, i.e., networked systems and Internet applications. This chapter first presents the *dominant state analysis* technique for modeling the inherent structure of data communications and characterizing the interaction of features within traffic clusters of individual networked systems and Internet applications. In addition, this chapter examines additional cluster features to provide further support and interpretation of network behavior analysis.

4.1 Communication Structure Analysis

4.1.1 Dominant State Analysis

The idea of dominant state analysis originates from structural modeling or reconstructability analysis in system theory [3, 4] as well as more recent graphical models in statistical learning theory [1]. The intuition behind the dominant state analysis is described below. Given a cluster, say a `srcIP` cluster, all flows in the cluster can be represented as a 4-tuple (ignoring the protocol field) $\langle u, x_i, y_i, z_i \rangle$, where the `srcIP` has a fixed value u, while the `srcPrt` (X dimension), `dsrPrt` (Y dimension), and `dstIP` (Z dimension) may take any legitimate values. Hence, each flow in the cluster imposes a "constraint" on the three "free" dimensions X, Y, and Z, as illustrated in Table 3.3 Treating each dimension as a random variable, the flows in the cluster constrain how the random variables X, Y, and Z "interact" or "depend" on each other, via the (induced) *joint* probability distribution $\mathcal{P}(X, Y, Z)$. The objective of dominant state analysis is to explore the interaction or dependence among the free dimensions by identifying "simpler" subsets of values or constraints (called *structural models* in the literature [2]) to represent or approximate the original data in their probability distribution. We refer to these subsets as *dominant states* of a traffic

K. Xu, *Network Behavior Analysis*,
https://doi.org/10.1007/978-981-16-8325-1_4

cluster. Hence, given the information about the dominant states, we can reproduce the original distribution with reasonable accuracy.

We use some examples to illustrate the basic ideas and usefulness of dominant state analysis. Suppose we have a `srcIP` cluster consisting mostly of scans (with a fixed `srcPrt` 220) to a large number of random destinations on `dstPrt` 6129. Then the values in the `srcPrt`, `dstPrt`, and `dstIP` dimensions these flows take are of the form $\langle 220, 6129, *\rangle$, where $*$ (wildcard) indicates random or arbitrary values. Clearly, this cluster belongs to `srcIP` BC_2 [0, 0, 2], and the cluster is dominated by the flows of the form $\langle 220, 6129, *\rangle$. Hence, the dominant state of the cluster is $\langle 220, 6129, *\rangle$, which approximately represents the nature of the flows in the cluster, even though there might be a small fraction of flows with other states. As a slightly more complicated example, consider a `srcIP` cluster which consists mostly of scanning traffic from the source (with randomly selected `srcPrt`) to a large number of random destinations on either `dstPrt` 139 (50% of the flows) or 445 (45%). Then the dominant states of the cluster (belonging to BC_{20}) are $\{\langle *, 139, *\rangle[50\%], \langle *, 445, *\rangle[45\%]\}$, where $[\cdot]$ indicates the percentage of flows captured by the corresponding dominant state. To emphasize the "interaction" or "dependence" among the dimensions, we represent the dominant states in the following form: {`dstPrt`(139) → (`srcPrt`(∗), `dstIP`(∗))[0.5], `dstPrt`(445) → (`srcPrt`(∗), `dstIP`(∗))[0.45]}, sometimes in short {`dstPrt`(139)[0.5], `dstPrt`} {(445)[0.45]}. Intuitively, it says that the `dstPrt` dimension determines the `srcPrt` and `dstIP` dimensions: by choosing 139 with probability 0.5 and 445 with probability 0.45 for the `dstPrt`, and then picking random values for the `srcPrt` and `dstIP`, we can reproduce a cluster that closely approximates the original cluster in terms of joint feature value distribution.

Figure 4.1 depicts the general procedure of dominant state analysis to extract dominant states from a cluster. Let $\{A, B, C\}$ be a re-ordering of the three free dimensions X, Y, Z of the cluster based on their RU values: A is the free dimension with the lowest RU, B the second lowest, and C the highest; in case of a tie, X always precedes Y or Z, and Y precedes Z. The dominant state analysis procedure starts by finding substantial values in the dimension A (step 1). A specific value a in the dimension A is substantial if the marginal probability

$$p(a) := \sum_b \sum_c p(a, b, c) \geq \delta, \tag{4.1}$$

where δ is a threshold for selecting substantial values. If no such substantial value exists, we stop. Otherwise, we proceed to step 2 and explore the "dependence" between the dimension A and dimension B by computing the conditional (marginal) probability, $p(b_j|a_i)$, of observing a value b_j in the dimension B given a_i in the dimension A:

$$p(b_j|a_i) := \sum_c \frac{p(a_i, b_j, c)}{p(a_i)}. \tag{4.2}$$

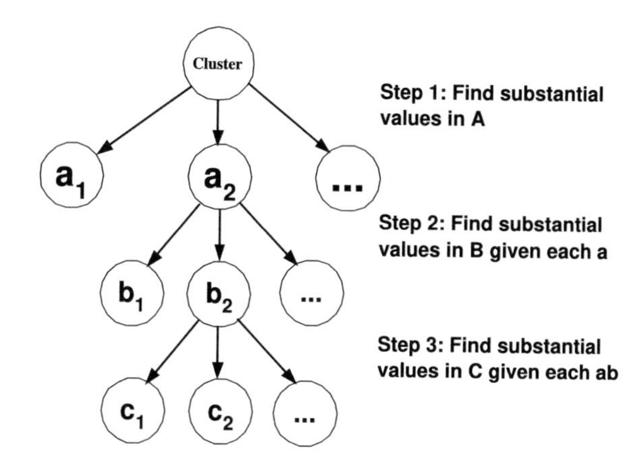

Fig. 4.1 General procedure for dominant state analysis

We find those substantial b_j's such that $p(b_j|a_i) \geq \delta$. If no substantial value exists, the procedure stops. Otherwise, we proceed to step 3 compute the conditional probability, $p(c_k|a_i, b_j)$, for each a_i, b_j and find those substantial c_k's, such that $p(c_k|a_i, b_j) \geq \delta$. The dominant state analysis procedure produces a set of dominate states of the following forms: $(*, *, *)$ (i.e., no dominant states), or $a_i \rightarrow (*, *)$ (by step 1), $a_i \rightarrow b_j \rightarrow *$ (by step 2), or $a_i \rightarrow b_j \rightarrow c_k$ (by step 3). The set of dominate states is an approximate summary of the flows in the cluster, and in a sense captures the "most information" of the cluster. In other words, the set of dominant states of a cluster provide a compact representation of the cluster.

4.1.2 Communication Structure of Networked Systems and Internet Applications

We apply the dominant state analysis to the clusters of four feature dimensions extracted on all links with varying δ in $[0.1, 0.3]$. The results with various δ are very similar, since the data is amenable to compact dominant state models. Table 4.1 (ignoring columns 4–7 for the moment, which we will discuss in the next section) shows dominant states of `srcIP` clusters extracted from link L_1 over a 1-hour period using $\delta = 0.2$. For each BC, the first row gives the total number of clusters belonging to the BC during the 1-hour period (column 2) and the general or prevailing form of the structural models (column 3) for the clusters. The subsequent rows detail the specific structural models shared by subsets of clusters and their respective numbers. The notations `dstIP`$(\cdot)$, `srcPrt`$(\cdots)$, etc., indicate a specific value and multiple values (e.g., in `dstIP`) that are omitted for clarity, and [>90%] denotes that the structural model captures at least 90% of the flows in the cluster (to avoid too much clutter in the table, this information is only shown for clusters in BC_2). The last column provides brief comments on the likely nature of the flows the clusters contain, which will be analyzed in more depth in next chapter.

Table 4.1 Dominant states for `srcIP` clusters on L_1 in a 1-hour period: $\delta = 0.2$

`srcIP` BC's	No. of clusters	Structural models	Range of $\mu(PKT)$	Range of $CV(PKT)$	Range of $\mu(BT)$	Range of $CV(BT)$	Brief comments
BC_2 [0, 0, 2]	119	rcPrt(·) → dstPrt(·) → dstIP(∗)	Small	Low	Small	Low	Mostly ICMP or scanning traffic
	114	srcPrt(0)→dstPrt(0)→dstIP(*)[>99%]	[1, 2]	[0, 1.6]	[72, 92]	[0, 8.9]	ICMP traffic
	1	srcPrt(1026)→dstPrt(137)→dstIP(*)[100%]	1	0	78	0	137: NetBIOS
	1	srcPrt(1153)→dstPrt(1434)→dstIP(*)[>98%]	1	0	404	0	1434: MS SQL
	3	srcPrt(220)→dstPrt(6129)→dstIP(*)[100%]	[1, 2]	[0, 1.2]	[40, 80]	[0, 2.6]	6129: Dameware
BC_6 [0, 2, 0]	16	srcPrt(·)→dstIP(· · ·)→dstPrt(*)	Large	High	Large	High	Server replying to a few hosts
	2	srcPrt(25)→dstIP(· · ·)→dstPrt(*)	[10, 15]	[1041, 2217]	[120, 750]	[36, 102]	25: Email
	5	srcPrt(53)→dstIP(· · ·)→dstPrt(*)	[1, 5]	[8.6, 78]	[160, 380]	[111, 328]	53: DNS
	7	srcPrt(80)→dstIP(· · ·)→dstPrt(*)	[3, 31]	$[460, 1.2 * 10^4]$	$[195, 1.2 * 10^5]$	[16, 1612]	80: Web
	2	srcPrt(443)→dstIP(· · ·)→dstPrt(*)	[3, 12]	$[320, 1.5 * 10^4]$	$[2166, 1.1 * 10^5]$	[29, 872]	443: https
BC_7 [0, 2, 1]	19	srcPrt(·)→dstIP(· · ·)→ dstPrt(*)	Large	High	Large	High	Server replying to many hosts
	2	srcPrt(25)→dstIP→dstPrt(*)	[14, 35]	[1129, 1381]]	[2498, 3167]]	[190, 640]	25: Email
	17	srcPrt(80)→dstIP→dstPrt(*)	[4, 26]	[210, 9146]	$[671, 1.0 * 10^4]$	[29, 3210]	80: Web
BC_8 [0, 2, 2]	7	srcPrt(.)→(dstPrt(*),dstIP(*))	Large	High	Large	High	Server replying to large # of hosts
	7	srcPrt(80)→(dstPrt(*),dstIP(*))	[4, 27]	$[1282, 1.1 * 10^4]$	$[740, 1.5 * 10^4]$	[72, 598]	80: Web
BC_{18} [2, 0, 0]	10	dstPrt(·)→(·)dstIP→srcPrt(*)	Medium	High	Medium	High	Host talking to a server on fixed `dstPrt`
	3	dstPrt(53)→dstIP→srcPrt(*)	[2, 5]	$[32, 1.5 * 10^5]$	[120, 325]	[82, 878]	53: DNS
	7	dstPrt(80)→dstIP→srcPrt(*)	[3, 18]	[26, 6869]	[189, 1728]	[87, 5086]	80: Web

(continued)

Table 4.1 (continued)

srcIP BC's	No. of clusters	Structural models	Range of $\mu(PKT)$	Range of $CV(PKT)$	Range of $\mu(BT)$	Range of $CV(BT)$	Brief comments
BC_{19} [2, 0, 1]	6	dstPrt(·)→dstIP(*)→srcPrt(*)	Medium	High	Medium	High	Host talking to multiple hosts on fixed `dstPrt`
	2	dstPrt(53)→dstIP(*)→srcPrt(*)	[2, 6]	[28, 875]	[116, 380]	[112, 456]	53: DNS
	3	dstPrt(80)→dstIP(*)→srcPrt(*)	[4, 16]	[72.3356]	[220, 2145]	[122, 2124]	80: Web
	1	dstPrt(7070)→dstIP(*)→srcPrt(*)	3	462	288	261	7070: RealAudio
BC_{20} [2, 0, 2]	58	dstPrt(·)→(srcPrt(*),dstIP(*))	Small	Low	Small	Low	Host talking to large # hosts on fixed `dstPrt`
	44	dstPrt(135)→(srcPrt(*),dstIP(*))	[1, 2]	[0, 1.6]	[48, 96]	[0, 2.7]	135: Microsoft RPC
	1	dstPrt(137)→(srcPrt(*),dstIP(*))	1	0	78	0	137: NETBIOS
	2	dstPrt(139)→(srcPrt(*),dstIP(*))	3	0	144	0	139: NETBIOS
	2	dstPrt(445)→(srcPrt(*),dstIP(*))	[1, 3]	[0, 2.2]	[48, 144]	[0, 3.6]	445: Microsoft-DS
	1	dstPrt(593)→(srcPrt(*),dstIP(*))	1	0	48	0	593: http RPC
	2	dstPrt(901)→(srcPrt(*),dstIP(*))	[1, 2]	[0, 1.6]	[48, 96]	[0, 3.9]	901: SMPNAMERES
	3	dstPrt(3127)→(srcPrt(*),dstIP(*))	[1, 3]	[0, 1.8]	[48, 144]	[0, 2.9]	3127: myDoom worm
	1	dstPrt(6129)→(srcPrt(*),dstIP(*))	1	0	40	0	6129: Dameware
	1	dstPrt(17300)→(srcPrt(*),dstIP(*))	1	0	48	0	17300: unknown
	1	dstPrt(34816)→(srcPrt(*),dstIP(*))	1	0.2	64	0.5	34816: unknown
BC_{24} [2, 2, 0]	1	dstIP(.)→srcPrt(*)→dstPrt(*)	–	–	–	–	Two hosts chatting on random ports
	1	dstIP(.)→srcPrt(*)→dstPrt(*)	1	0	889	0	vertical scan

The results in the table demonstrate two main points. First, clusters within a BC have *(nearly) identical* forms of structural models; they differ only in specific values they take. For example, BC_2 and BC_{20} consist mostly of hosts engaging in various scanning or worm activities using known exploits, while `srcIP` clusters in BC_6, BC_7, and BC_8 are servers providing well-known services. They further support our assertion that the RU-based behavior classification scheme, presented in Chap. 3, automatically groups together clusters with similar behavior patterns, despite that the classification is done *oblivious* of specific feature values that flows in the clusters take. Second, the structural model of a cluster presents a compact summary of its constituent flows by revealing the essential information about the cluster (substance feature values and interaction among the free dimensions). It in itself is useful, as it provides *interpretive value* to network operators for understanding the cluster behavior. These points also hold for clusters extracted from other dimensions. For instance, Table 4.2 presents the summarized results obtained for `dstPrt` clusters extracted from the same 1-hour period on L_2.

4.2 Exploring More Traffic Features

So far, we have focused the analysis on the four key feature dimensions, `srcIP`, `dstIP`, `srcPrt`, and `dstPrt`, and use RU measures along these dimensions to automatically classify significant cluster behaviors—that the behavior classes indeed characterize clusters of similar behavior patterns is corroborated by the dominant state analysis presented above. We now investigate whether additional features (beyond the four basic features, `srcIP`, `dstIP`, `srcPrt`, and `dstPrt`) can (i) provide further affirmation of similarities among clusters within a BC, and in case of wide diversity, (ii) be used to distinguish sub-classes of behaviors within a BC. Examples of additional features we consider are cluster sizes (defined in total flow, packet, and byte counts), average packet/byte count per flow within a cluster and their variability, etc. In the following, we illustrate the results of additional feature exploration using the average flow sizes per cluster and their variability.

For each flow f_i, $1 \le i \le m$, in a cluster, let PKT_i and BT_i denote the number of packets and bytes, respectively, in the flow. The average number of packets for the cluster, $\mu(PKT)$, is computed as

$$\mu(PKT) = \frac{\sum_i PKT_i}{m}, \tag{4.3}$$

while the average number of bytes for the cluster, $\mu(BT)$, is computed as

$$\mu(BT) = \frac{\sum_i BT_i}{m}. \tag{4.4}$$

Table 4.2 Dominant states for `dstPrt` clusters on L_1 in a 1-hour period: $\delta = 0.2$

`dstPrt` BC's	No. of clusters	Structural models	Range of $\mu(PKT)$	Range of $CV(PKT)$	Range of $\mu(BT)$	Range of $CV(BT)$	Brief comments (Possible interpretation)
BC_2 [0, 0, 2]	1	srcPrt(220)→srcIP(...)→dstIP(*)	[1, 1.7]	[0, 5.9]	[40, 68]	[0, 12]	A few hosts exploit port 6129 on random dstIPs
BC_5 [0, 1, 2]	1	srcPrt(0)→srcIP(...)→dstIP(*)	[1.2, 6]	[510, 900]	[120, 670]	[112, 492]	Many hosts send ICMP to random dstIPs
BC_{15} [1, 2, 0]	1	dstIP(.)→srcPrt(*)→srcIP(*)	1	1.4	40.19	0.04	dstPrt(6667), DDoS attack on a particular dstIP
BC_{20} [2, 0, 2]	2	srcIP(...)→srcPrt(*)→dstIP(*)	[1, 3]	[0, 2.1]	[40, 160]	[0, 7.9]	dstPrt(135/137) exploit traffic
BC_{22} [2, 1, 0]	1	dstIP(.)→srcIP(...)→srcPrt(*)	[12, 21]	[2500, 3700]	[3024, 9800]	[180, 267]	dstPrt(443) https traffic
BC_{23} [2, 1, 2]	12	srcIP(...)→srcPrt(*)→dstIP(*)	[1, 2]	[0, 1.39]	[50, 132]	[0, 9]	dstPrt135/137/139/445/593/901/1433/1434/3127/6129/12345/34816) exploit traffic
BC_{25} [2, 2, 1]	6	dstIP(...)→srcPrt(*)→srcIP(*)	[4, 25]	[2711, 5138]	[402, 15600]	[98, 270]	dstPrt(25/53/80), dst-Prt(1214/4662/6346) service traffic, and P2P traffic

We also measure the flow size variability in packets, $CV(PKT)$, using *coefficient of variance*

$$CV(PKT) = \frac{\sigma(PKT)}{\mu(PKT)} \tag{4.5}$$

and measure the flow size variability in bytes, $CV(BT)$, which is calculated as

$$CV(BT) = \frac{\sigma(BT)}{\mu(BT)}, \tag{4.6}$$

where $\sigma(PKT)$ and $\sigma(BT)$ are the standard deviation of PKT_i and BT_i.

In Table 4.1, columns 4–7, we present the ranges of $\mu(PKT)$, $CV(PKT)$, $\mu(BT)$, and $CV(BT)$ of subsets of clusters with the similar dominant states, using the 1-hour `srcIP` clusters on L_1. Columns 4–7 in the top row of each BC are high-level summaries for clusters within a BC (if it contains more than one cluster): small, medium, or large average packet/byte count, and low or high variability. We see that for clusters within BC_6, BC_7, BC_8, and BC_{18}, BC_{19}, the average flow size in packets and bytes are at least 5 packets and 320 bytes, and their variabilities ($CV(PKT)$ and $CV(BT)$) are fairly high. In contrast, clusters in BC_2 and BC_{20} have small average flow size with low variability, suggesting most of the flows contain a singleton packet with a small payload. The same can be said of most of the less popular and rare BCs.

Finally, Fig. 4.2a–d show the average cluster sizes in flow, packet and byte counts for all the unique clusters from the dataset L_1 within four different groups of BC's: $\{BC_6, BC_7, BC_8\}$, $\{BC_{18}, BC_{19}\}$, $\{BC_2, BC_{20}\}$, and the fourth group containing the remaining less popular BC's. Note that the average sizes for clusters are computed only for clusters appearing twice or more. If a cluster is observed only once, the average cluster sizes will be the same as the actual flow, packet, and byte counts Clearly, the characteristics of the cluster sizes of the first two BC groups are quite different from those of the second two BC groups. Such differences could potentially provide valuable insights into anomaly detection. To conclude, our results demonstrate that BC's with distinct behaviors, e.g., non-akin BC's, often also manifest dissimilarities in other features. Clusters within a BC may also exhibit some diversity in additional features, but in general the intra-BC differences are much less pronounced than inter-BC differences.

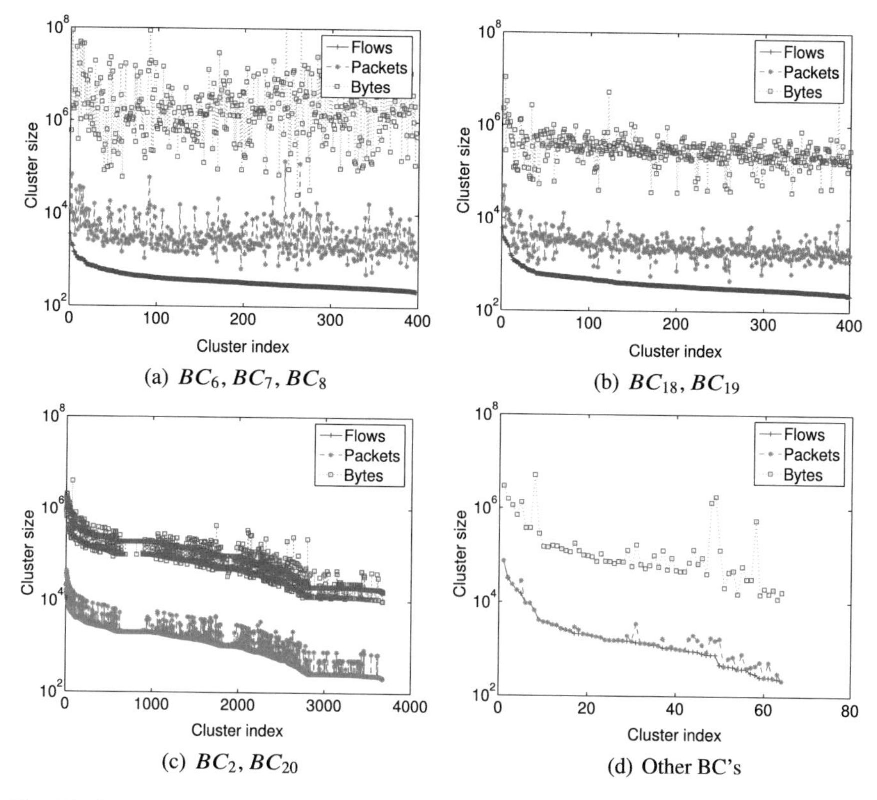

(a) BC_6, BC_7, BC_8 (b) BC_{18}, BC_{19} (c) BC_2, BC_{20} (d) Other BC's

Fig. 4.2 Average cluster size (in flow, packet, and byte count) distributions for clusters within four groups of BC's for srcIP clusters on L_1. Note that in **c** and **d**, the lines of flow count and packet count are indistinguishable, since most flows in the clusters contain a singleton packet

4.3 Summary

In summary, the dominant state analysis and additional feature inspection have collectively provided plausible interpretation of behavioral patterns of networked systems and Internet applications. Such a compact structural model for traffic clusters based on dominant states captures the most common or significant feature values and their interaction. In other words, dominant state analysis is an intuitive and effective technique for modeling communication structures of network traffic characterizing the interaction of features within traffic clusters, and capturing similarities/dissimilarities among behavior classes and individual clusters.

The dominant state analysis provides two critical contributions in network behavior analysis. First, it provides support for the behavior classification—we find that clusters within a behavior class have nearly identical forms of structural models. Second, it yields compact summaries of cluster information which provides interpretive value to network operators for explaining observed behavior, and helps in narrowing down the scope of a deeper investigation into specific clusters.

References

1. M. Jordan, Graphical models. Stat. Sci. Special Issue on Bayesian Stat. **19**, 140–155 (2004)
2. K. Krippendorff, *Information Theory: Structural Models for Qualitative Data* (Sage Publications, Newbury Park, 1986)
3. M. Zwick, An overview of reconstructability analysis. Int. J. Syst. Cybern. (2004)
4. R. Cavallo, G. Klir, Reconstructability analysis of multi-dimensional relations: a theoretical basis for computer-aided determination of acceptable systems models. Int. J. Gen. Syst. **5**, 143–171 (1979)

Chapter 5
Graphical Modeling of Network Traffic

Abstract As networked systems and Internet applications continue to grow, it becomes increasingly important to understand their traffic patterns for efficient network management and security monitoring. A number of research studies have focused on traffic behavior analysis of individual hosts and applications. However, an increasingly large number of networked systems, a wide diversity of applications, and massive traffic data pose significant challenges for such fine-granularity analysis for backbone networks or enterprise networks. This chapter presents a graphical approach to profiling traffic behavior by identifying and analyzing *clusters* of hosts or applications that exhibit similar communication patterns. With each cluster abstracting behavior patterns of a plurality of hosts or applications, the cost of traffic analysis is significantly reduced. This chapter first explains the rationale, importance, benefits, and challenges of performing cluster-aware network behavior analysis. Subsequently, this chapter discusses how to explore bipartite graphs for modeling data communication in network traffic and the one-mode projection for capturing behavior similarity of networked systems and describes the similarity matrices and clustering coefficient of one-mode projection graphs. The availability of similarity matrices motivates the usage of clustering algorithms to leverage similarity matrices and clustering coefficient for discovering behavior clusters of networked systems in the same prefixes or engaging in the same applications. Finally, this chapter presents the distinct traffic characteristics of end-host behavior clusters within the same network prefixes and explores the behavior similarity of Internet applications.

5.1 Cluster-Aware Network Behavior Analysis

As Internet hosts and applications continue to grow, it becomes increasingly important to understand traffic patterns of networked systems and Internet applications for efficient network management and security monitoring. A number of research studies [1–4] have focused on traffic behavior analysis of individual systems and applications. However, an increasingly large number of networked systems, a wide

K. Xu, *Network Behavior Analysis*,
https://doi.org/10.1007/978-981-16-8325-1_5

diversity of applications, and massive traffic data pose significant challenges for such fine-granularity analysis for backbone networks or enterprise networks.

This chapter introduces a new approach of profiling traffic behavior by identifying and analyzing clusters of hosts or applications that exhibit similar communication patterns. With each cluster abstracting behavior patterns of a plurality of hosts or applications, the cost of traffic analysis is significantly reduced. We first use bipartite graphs to model network traffic of Internet backbone links or Internet-facing links of border routers in enterprise networks. As one-mode projections can effectively extract hidden relationships between nodes within the same vertex sets of bipartite graphs [5], we subsequently construct one-mode projections of bipartite graphs to connect source hosts that communicate with the same destination host(s), and to connect destination hosts that communicate with the same source host(s).

The derived one-mode projection graphs enable us to further build similarity matrices of networked systems, with similarity being characterized by the shared number of destinations or sources between two hosts. Based on the similarity matrices of networked systems in the same network prefixes, we apply a simple yet effective spectral clustering algorithm to discover the inherent *end-host behavior clusters*. Each cluster consists of a group of hosts that communicate with similar sets of servers, clients, or peers. The behavior clusters not only reduce the number of behavior profiles for analysis compared with traffic profiling on individual hosts, but also reveal detailed behavior patterns for a group of networked systems in the same network prefixes.

Similarly, we use a vector of graph properties including clustering coefficient to capture the similarity of traffic behavior for networked systems engaging in the same Internet applications, and discover the inherent *application behavior clusters*, each of which consists of a number of applications. For each application cluster, we examine characteristics of the aggregated traffic, such as host symmetry, the fan-out degree of source IP addresses, and the fan-in degree of destination IP addresses. The experimental results based on real Internet backbone traffic confirm that application behavior clusters indeed capture applications with similar traffic characteristics and behavior patterns.

5.2 Modeling Host Communications with Bipartite Graphs and One-Mode Projections

Data communications observed unidirectional Internet links can naturally be represented by bipartite graphs where all IP packets originate from one set of nodes, i.e., source IP addresses, to another disjoint set of nodes, i.e., destination IP addresses. Let G represent the bipartite graph to model such data communications $\mathcal{G} = (\mathcal{A}, \mathcal{B}, \mathcal{E})$, where $\mathcal{A}$ and $\mathcal{B}$ represent the disjoint vertex sets of source and destination IP addresses in the graph, and $\mathcal{E} \subseteq \mathcal{A} \times \mathcal{B}$ is the edge set in the bipartite graph $\mathcal{G}$.

To analyze the traffic behavior for network prefixes which include networked systems with the same network bits in their IP addresses, we could further decompose the bipartite graph of all the traffic into a set of smaller disjoint bipartite subgraphs such that each bipartite subgraph captures the host communications for a single source or destination IP prefix, e.g., *source host behavior graph (SHBG)* $\mathcal{G}_{\mathcal{P}} = (\mathcal{A}_{\mathcal{P}}, \mathcal{B}, \mathcal{E}_{\mathcal{P}})$ and *destination host behavior graph (DHBG)* $\mathcal{G}_{\mathcal{Q}} = (\mathcal{A}, \mathcal{B}_{\mathcal{Q}}, \mathcal{E}_{\mathcal{Q}})$ representing the bipartite subgraphs of host communications for the source IP prefix $\mathcal{P}$ and the destination IP prefix $\mathcal{Q}$, respectively.

Similarly, for a given application port, its traffic also forms a natural subgraph of the bipartite graph. Let $\mathcal{A}_{\texttt{port\_number}}$ and $\mathcal{B}_{\texttt{port\_number}}$ denote the sets of source and destination IP addresses engaging in the application port `port_number`, respectively. Then we could build two bipartite subgraphs *source port behavior graph (SPBG)* $\mathcal{G}_{\texttt{srcport}} = (\mathcal{A}_{\texttt{srcport}}, \mathcal{B}, \mathcal{E}_{\texttt{srcport}})$ and *destination port behavior graph (DPBG)* $\mathcal{G}_{\texttt{dstport}} = (\mathcal{A}, \mathcal{B}_{\texttt{dstport}}, \mathcal{E}_{\texttt{dstport}})$ for representing host communications for the source port `srcport` and the destination port `dstport`, respectively.

The one-mode projection of the bipartite graphs uses edges between networked systems in the same network prefixes or engaging in the same application to quantify the similarity of their network connection patterns. For example, in Fig. 2.2, the edge between s_1 and s_2 reflects the observation that both s_1 and s_2 talk with the same destination host d_1 in the bipartite graph (Fig. 2.2), and the edge between d_2 and d_3 in Fig. 2.3b captures the observation that s_3 talks with both destinations d_2 and d_3. Therefore, given a bipartite graph $\mathcal{G}_{\mathcal{P}} = (\mathcal{A}_{\mathcal{P}}, \mathcal{B}, \mathcal{E}_{\mathcal{P}})$ for a source prefix $\mathcal{P}$, we could construct the one-mode projection graph of *SHBG on source prefix* $\mathcal{P}$, $\mathcal{G}'_{\mathcal{A}_{\mathcal{P}}} = (\mathcal{A}_{\mathcal{P}}, \mathcal{E}'_{\mathcal{A}_{\mathcal{P}}})$, where $\mathcal{A}_{\mathcal{P}}$ consists of all source hosts observed in $\mathcal{P}$ and $\{p_i, p_j\} \in \mathcal{E}'_{\mathcal{A}_{\mathcal{P}}}$ if and only if two hosts p_i and p_j talk with at least one same destination host. The similar process could generate the one-mode projection graph of *DHBG on destination prefix* $\mathcal{Q}$ for any destination prefix $\mathcal{Q}$ as well. Using the same approach, we could build one-mode projection graphs of *SPBG on port* `srcport` and of *DPBG on port* `dstport`. In this study, we leverage one-mode projection graphs to explore the social-behavior similarity of source or destination IP addresses that share the same network prefixes or engage in the same Internet applications.

5.3 Similarity Matrices and Clustering Coefficient of One-Mode Projection Graphs

5.3.1 Similarity Matrices

To capture the information on the degree of the social-behavior similarity among networked systems, we use the normalized weight for the edges in the one-mode projection graph. Let $\mathcal{N}_{p_i}$ and $\mathcal{N}_{p_j}$ represent the numbers of Internet hosts which two hosts p_i and p_j in the prefix $\mathcal{P}$ have communicated with, respectively. We then use $w_{\{p_i, p_j\}}$ to denote the weight for the edge between p_i and p_j in the one-mode

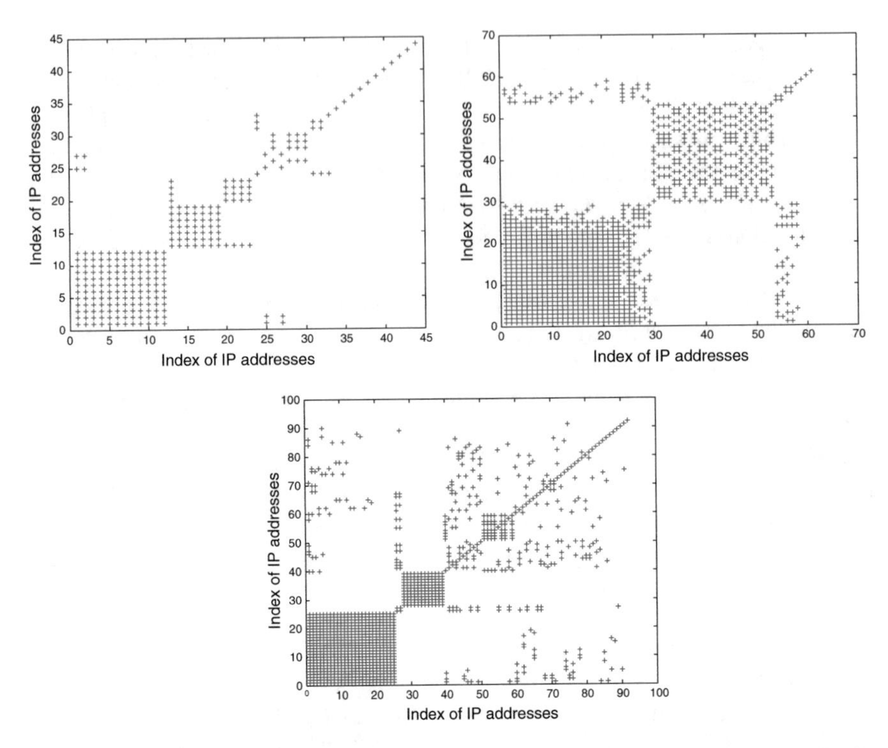

Fig. 5.1 Visualization of the adjacency matrix for one-mode projections of bipartite graphs for three network prefixes

projection,

$$w_{\{p_i,p_j\}} = \frac{|\mathcal{N}_{p_i} \bigcap \mathcal{N}_{p_j}|}{|\mathcal{N}_{p_i} \bigcup \mathcal{N}_{p_j}|}, \tag{5.1}$$

where $|\mathcal{N}_{p_i} \bigcap \mathcal{N}_{p_j}|$ denotes the total number of the shared destination hosts in the bipartite graph between the two hosts p_i and p_j, and $|\mathcal{N}_{p_i} \bigcup \mathcal{N}_{p_j}|$ denotes the total number of the uniquely combined destinations of p_i and p_j. Note that $w_{\{p_i,p_i\}} = 1$. The weighted adjacency matrix of the one-mode projection graph for the network prefix $\mathcal{P}$ then becomes $\mathcal{M}_{\mathcal{P}} = (m_{i,j})_{|\mathcal{P}|\times|\mathcal{P}|}$, $m(i,j) = w_{\{p_i,p_j\}}$. The similar process could lead to the weighted adjacency matrices $\mathcal{M}_{\mathcal{Q}}$, $\mathcal{M}_{\texttt{srcport}}$, and $\mathcal{M}_{\texttt{dstport}}$ of the one-mode projection graph for the destination prefix $\mathcal{Q}$, the source port `srcport`, and the destination port `dstport`, respectively.

One interesting observation of the one-mode projection graphs for host communications lies in the clustered patterns in the weighted adjacency matrix. The scatter plots in Fig. 5.1 visualize the adjacency matrices of the one-mode projection graphs for three different network prefixes with 44, 61, and 92 networked systems, respectively. For each prefix, we sort the IP addresses based on the hosts' degree (number of neighbors in the one-mode project graph) in a non-increasing order. Both x-axis and y-axis represent the indices of IP addresses in the same prefix, and each "+" point (i, j) in the plots denotes an edge with a positive weight between two sorted hosts p_i and p_j in the one-mode projection graph, i.e., $m(i, j) = w_{\{p_i, p_j\}} > 0$. As shown in Fig. 5.1, each prefix has a few well-separated blocks that divide networked systems into different clusters. This observation on the adjacency matrix leads to the next step of further exploring cluster analysis techniques and graph partitioning algorithm [6] to uncover these behavior clusters of networked systems that share the same network prefixes or engage in the same Internet applications.

5.3.2 Clustering Coefficients

Clustering coefficient is a widely used measure to study the "closeness", or the "small-world" patterns of nodes in one-mode projection graphs [7]. This measure can be applied to individual nodes as *local clustering coefficient (LCC)* and can also be applied to the entire graph as *global clustering coefficient (GCC)*. For a given node u, the local clustering coefficient, LCC_u, is provided by the number of the edges among u's neighbors over the number of all possible edges among u's neighbors. Let N_u represent the set of all the neighbors of the node u, where $|N_u| = m$, and let E_u represent the set of edges among these neighbors. The number of all possible edges among m neighbors is $\frac{m\times(m-1)}{2}$. The *local clustering coefficient (LCC)* of u is calculated as follows:

$$LCC_u = \frac{|E_u|}{(m * (m - 1))/2} = \frac{|E_u| * 2}{m * (m - 1)}. \tag{5.2}$$

Clearly $LCC_u \in [0, 1]$. LCC_u is 0 if there is no edge among u's neighbors, while LCC_u is 1 if u's neighbors form a complete graph (clique). Note that the *local clustering coefficient (LCC)* for nodes with 0 or 1 neighbor is 0 due to zero edges. The *global clustering coefficient (GCC)* of the entire graph, GCC_G, is the average *local clustering coefficient (LCC)* over all n nodes, where

$$GCC_G = \frac{1}{N} * \sum LCC_u, \tag{5.3}$$

where $u \in G$. Because of the existence of nodes with 0 or 1 neighbor which affect the calculation of the global clustering coefficient, we adopt an adaptive global clustering coefficient (AGCC), introduced in [8],

$$AGCC_G = \frac{1}{1 - \theta} * GCC_G, \tag{5.4}$$

where θ is the percentage of the isolated nodes in one-mode projection graphs. In addition, we also measure the percentage of the nodes that have at least two neighbors (or non-isolated nodes) in the graphs. In Sect. 5.5.3, we will show how clustering

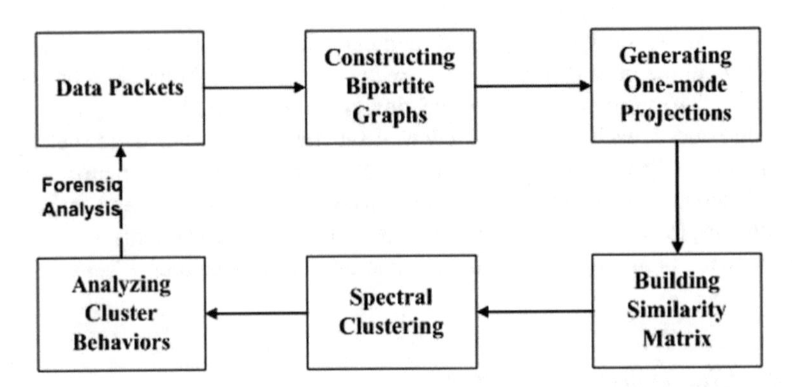

Fig. 5.2 The schematic process of network-aware behavior clustering algorithm for discovering behavior clusters of network prefixes

coefficient captures social behavior of source or destination IP addresses engaging in the same applications with real network traffic datasets and helps uncover groups of Internet applications exhibiting similar traffic patterns.

5.4 Discovering Behavior Clusters via Clustering Algorithms

5.4.1 Partitioning Similarity Matrix with Spectral Clustering Algorithm

Clustering algorithms have been used to analyze and profile hosts on the Internet. For example, the study in [3] uses an agglomerative clustering algorithm to characterize networked systems based on traffic features in IP packet headers, such as the number of distinct destination IP addresses, the daily count of network traffic volumes in bytes, average TTL (time-to-live) value, etc. In the study of graphical modeling of network traffic, we focus on the social behavior of networked systems in data communications through bipartite graphs and one-mode projection graphs, and are interested in exploring the social-behavior similarity of networked systems to discover inherent traffic clusters.

Figure 5.2 illustrates the schematic process of our clustering approach from constructing bipartite graphs based on IP packets to discovering and analyzing behavior clusters of network prefixes.

An important starting point of a clustering algorithm is to define the appropriate similarity matrix between data points. Here, we use the weighted edge between two hosts u and v of the same prefix in the one-mode projection graph as the similarity measure $s_{u,v}$ between u and v, because the weighted edges capture and quantify the social-behavior similarity of host communications in network traffic. Therefore, the weighted adjacency matrix of the one-mode projection graphs for the prefix $\mathcal{M}_{\mathcal{P}}$ essentially becomes the similarity matrix $\mathcal{S}_{\mathcal{P}}$ which will be used as an input to the spectral clustering algorithm outlined below.

This study applies a simple spectral clustering algorithm developed in [6] due to its wide applications in graph partitioning and its small running time. The original spectral clustering algorithm [6] requires an explicit input of k as the expected number of clusters. Given the infeasibility of predicting the optimal number of behavior clusters in network prefixes without analyzing the traffic data, we therefore augment the algorithm by adding a step of automatically selecting an appropriate value of k as the desired number of the clusters based on the eigenvalue distribution. The detail of this step is explained in the following algorithm:

Algorithm 2 Algorithm of discovering behavior clusters using an augmented spectral clustering algorithm

---

Input: network flow traces during a given time window and a source or destination prefix $\mathcal{P}$;

1: Construct bipartite graphs of host communications from flow traces;

2: Generate the one-mode projection of bipartite graphs and its weighted adjacency matrix $\mathcal{M}_{\mathcal{P}}$ for networked systems in the prefix $\mathcal{P}$, and then obtain the similarity matrix $\mathcal{S}_{\mathcal{P}} \in \mathbb{R}^{n \times n}$ for the prefix $\mathcal{P}$;

3: Let A be the diagonal matrix with $A(i,i) = \sum_{j=1}^{n} s_{i,j}$, where $i = 1, \ldots, n$;

4: Compute the Laplacian matrix $L = A^{-1/2} S A^{-1/2}$;

5: Find the largest k eigenvalues, $\lambda_1, \lambda_2, \ldots, \lambda_k$ such that $\sum_{i=1}^{k} \lambda_i \geq \alpha \times \sum_{j=1}^{n} \lambda_n$ and $(\lambda_k - \lambda_{k+1}) \geq \beta \times (\lambda_{k-1} - \lambda_k)$;

6: Use the corresponding k eigenvectors $(e_1, e_2, \ldots, e_k)$ as columns to construct the matrix $E = [e_1 e_2 \ldots e_k] \in \mathbb{R}^{n \times k}$;

7: Construct the matrix $\mathcal{Z}$ through renormalizing E such that each row has a unit length, and consider each row as a point;

8: Run k-means clustering algorithm to cluster the points of $\mathcal{Z}$ into k clusters $(\mathcal{Y}_1, \mathcal{Y}_2, \ldots, \mathcal{Y}_k)$

9: Assign the original IP address p_i to the cluster C_j if the row i of $\mathcal{Z}$ is assigned to the cluster $\mathcal{Y}_j$.

Output: clusters $C_1, C_2, \ldots, C_k$, where $C_i = \{p_j | z_j \in \mathcal{Y}_j\}$.

Algorithm 1 outlines the major steps of the spectral clustering algorithm with the augmented change of automatically selecting k clusters based on the traffic patterns. The input of this algorithm is network flow traces during a given time window and a source or destination prefix $\mathcal{P}$. The first step is to use flow traces to construct bipartite graphs of host communications, while the second step is to generate the one-mode projection of bipartite graphs and its weighted adjacency matrix $\mathcal{M}_{\mathcal{P}}$ for networked systems in the prefix $\mathcal{P}$, and then to obtain the similarity matrix $\mathcal{S}_{\mathcal{P}} \in \mathbb{R}^{n \times n}$.

Next, we compute the Laplacian matrix $L = A^{-1/2} S A^{-1/2}$, where A is the diagonal matrix with $A(i,i) = \sum_{j=1}^{n} s_{i,j}$ and $i = 1, \ldots, n$. Then in the augmented step we search for the largest k eigenvalues, $\lambda_1, \lambda_2, \ldots, \lambda_k$ such that $\sum_{i=1}^{k} \lambda_i \geq \alpha \times \sum_{j=1}^{n} \lambda_n$ and $(\lambda_k - \lambda_{k+1}) \geq \beta \times (\lambda_{k-1} - \lambda_k)$. In other words, the augmented step searches an appropriate value for k by finding the largest k eigenvalues that account for at least α of the total variances and stopping at the eigenvalue λ_k where the distribution of

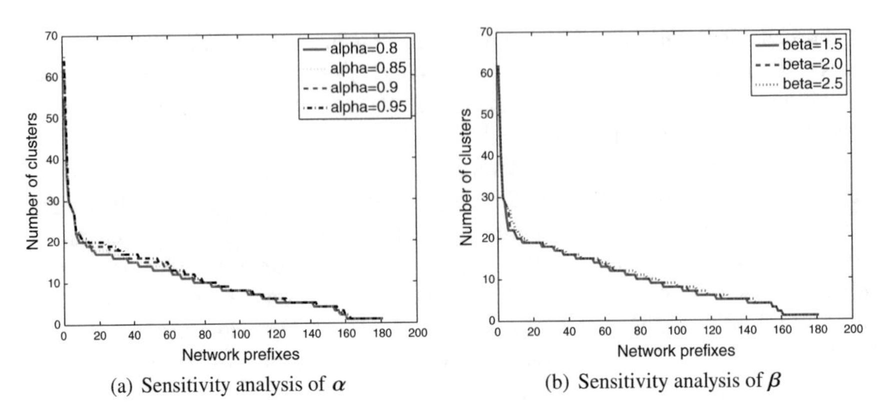

(a) Sensitivity analysis of α (b) Sensitivity analysis of β

Fig. 5.3 Sensitivity analysis of α and β used in the proposed algorithm of discovering behavior clusters

eigenvalues exhibits a sharp slope change. In our experiments, we have evaluated a variety of values for α and β, and found that there are no significant changes for α in the range of [0.8, 0.95] and β in the range of [1.5, 2.5]. For example, Fig. 5.3a shows the similar numbers of discovered clusters by the proposed algorithm for all source network prefixes during a 1-min time window with $\beta = 2$ and α being set as 0.8, 0.85, 0.09, and 0.95, respectively, while Fig. 5.3b also shows the similar numbers of clusters for the same set of network prefixes with $\alpha = 0.9$ and β being set as 1.5, 2.0, and 2.5, respectively. Thus, in the remaining of this chapter, we use 0.9 and 2 for the α and β, respectively, to present the experimental results.

In the experiments with real traffic traces, we find that it is common to observe that a few eigenvectors account for the majority of the variances in the similarity matrix for IP prefixes. Thus, we use the corresponding top k eigenvectors ($e_1, e_2, \ldots, e_k$) as columns to construct the matrix $E = [e_1 e_2 \ldots e_k] \in \mathbb{R}^{n\times k}$, and subsequently construct the matrix $\mathcal{Z}$ through renormalizing E such that each row has a unit length. Considering each row as a point, the final step of the algorithm is to run a k-means clustering algorithm to cluster the points of $\mathcal{Z}$ into k clusters ($\mathcal{Y}_1, \mathcal{Y}_2, \ldots, \mathcal{Y}_k$), and then assign the original IP address p_i to the cluster C_j if the row i of $\mathcal{Z}$ is assigned to the cluster $\mathcal{Y}_j$.

The output of this algorithm is a set of k clusters ($C_1, C_2, \ldots, C_k$), each of which includes a group of networked systems sharing similar social-behavior patterns in network traffic. In the following section, we will study traffic characteristics of end-host behavior clusters discovered by the spectral clustering algorithm, and then demonstrate the practical benefits of these clusters for discovering traffic patterns and detecting anomalous behaviors.

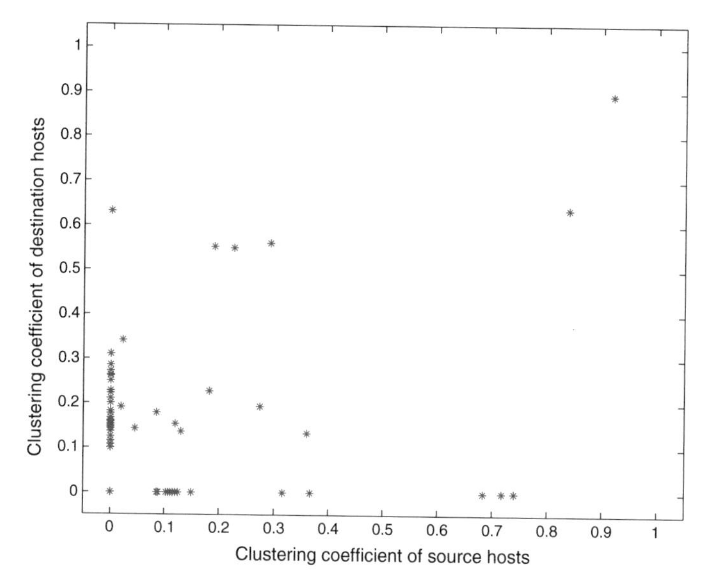

Fig. 5.4 Distribution of adaptive global clustering coefficient for source and destination behavior graphs of Internet applications

5.4.2 *Clustering Analysis of Internet Applications*

For the source and destination behavior graphs generated from the traffic of each Internet application, we calculate the *adaptive global clustering coefficient*. In this study, we consider a unique combination of port number and transport protocol (TCP or UDP) as one Internet application. For example, all network traffic on port 80/TCP is considered as an Internet application. In addition, we focus on network applications with consistent port numbers. Some applications, e.g., peer-to-peer file sharing that use random port numbers to obfuscate their traffic behavior, require additional information, e.g., packet payload and hosts with labeled traffic patterns, to study social behavior of source and destination hosts. In this book, we refer to the *adaptive global clustering coefficient* as *clustering coefficient* for simplicity. Figure 5.4 shows the distribution of clustering coefficient for all Internet applications observed from an OC192 Internet backbone link during a 1-min time window. An interesting observation is that the clustered pattern of clustering coefficient, which leads to our next step of applying clustering algorithms to discover the inherent clusters formed by Internet applications sharing similar behavior patterns.

Based on clustering coefficient and other graph properties of Internet applications, we apply a simple K-means clustering algorithm [9] to group them into distinct application behavior clusters. The choice of selecting this algorithm is due to its simplicity and wide usage. The features used in the clustering algorithm include clustering coefficients of source and destination behavior graphs, and the ratios of nodes with two or more neighbors in these graphs. In other words, for each source

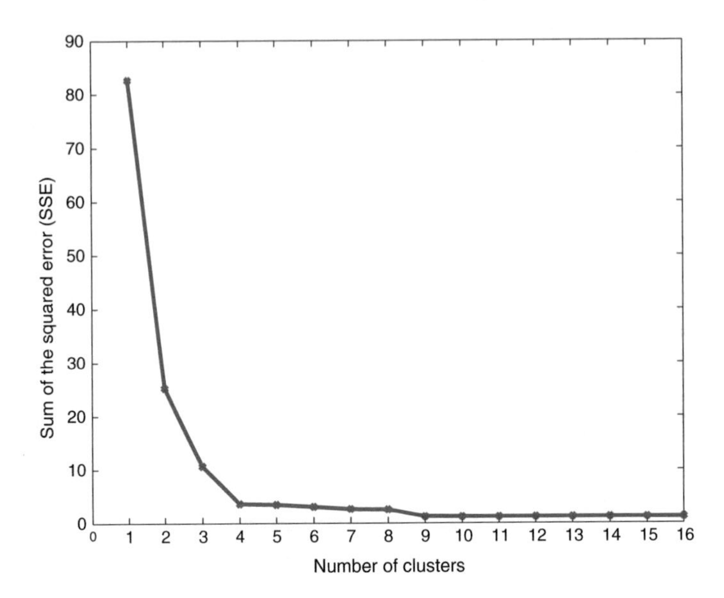

Fig. 5.5 Determining the optimal k based on sum of squared error (SSE)

or destination port p, we obtain a vector of four features, i.e., $AGCC_{\mathcal{S}_p}$, $AGCC_{\mathcal{D}_p}$, $r_{\mathcal{S}_p}$, and $r_{\mathcal{D}_p}$, where the first two features are clustering coefficients of source and destination hosts engaging in the application port p and the last two features are ratios of hosts with at least two or more neighbors in one-mode projection graphs on source and destination hosts.

A challenging issue of applying K-means clustering algorithms is to find an optimal value of k, since the choice of k plays an important role of archiving the high quality of clustering results. Towards this end, we search the optimal value of k by running K-means algorithms using a variety of k values and evaluate the best choice of k by comparing the sum of squared error (SSE) with Euclidean distance function between nodes in each cluster [9]. For example, Fig. 5.5 illustrates the distribution of SSE with varying values of k from 1 to 16. We select $k = 9$ as the choice since increasing k from 9 to 10 and above does not bring significant benefits of reducing SSE.

5.5 Traffic Characteristics and Similarity of Behavior Clusters

5.5.1 Making Sense of End-Host Behavior Clusters

5.5.1.1 Datasets

The datasets used in our analysis are collected from CAIDA's equinix-chicago and equinix-sanjose network monitors [10] on bidirectional OC192 Internet backbone links of a large Internet service provider during December 17, 2009. The CAIDA Internet traffic traces are anonymized using CryptoPAn *prefix-preserving* anonymization [11] for privacy reasons, however such *prefix-preserving* process does not affect our analysis that explores behavior similarity of networked systems within the same network prefixes or engaging in the same Internet applications.

Similar to the observations in previous studies [12], Internet links carry large volumes of network traffic, which poses a challenging problem for real-time or near-real-time traffic analysis. The total size of the compressed dataset used in this study is over 200 GB. As a first step to reduce the data size, we aggregate packet traces into

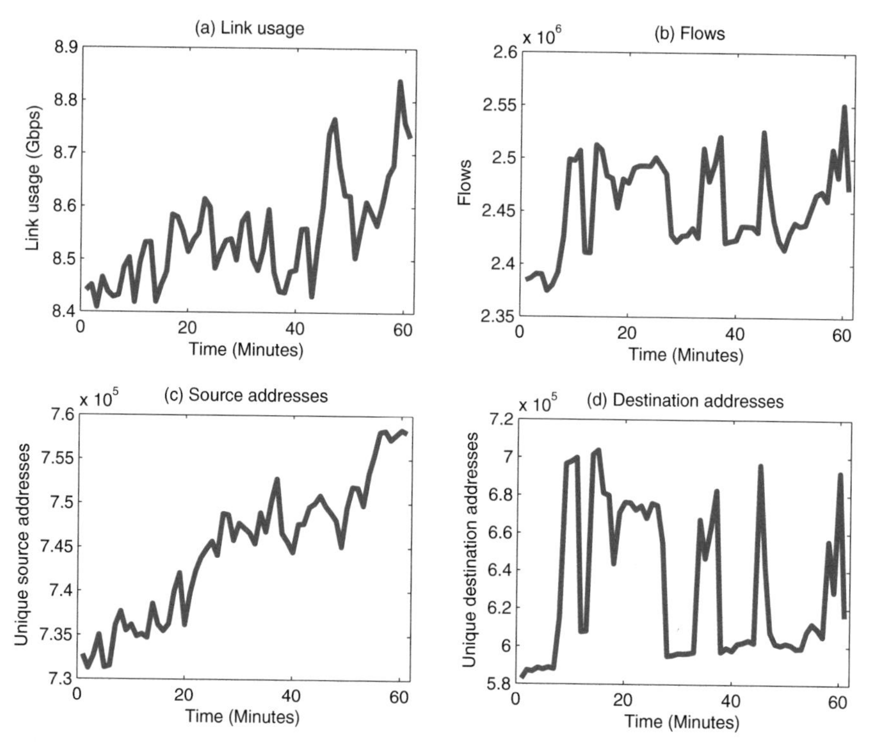

Fig. 5.6 Statistics of CAIDA Internet traffic trace

the well-known 5-tuple network flows. Figure 5.6a shows an average of 8.6 Gbps link usage during a 1-hour duration, and Fig. 5.6b illustrates millions of network flows for every minute during this period. In our analysis, we use 1-min time bin to analyze traffic data due to vast amounts of packets and flows to be processed in each time bin. In addition, Fig. 5.6c, d show the total numbers of unique source and destination IP addresses, respectively. Such a large number of unique IP addresses in the packet traces make it very challenging to analyze traffic behavior at host level [3], and therefore the focus on behavior clusters of network prefixes becomes an intuitive alternative for scalable analysis on Internet backbone traffic.

In our analysis, we use /24 block as the network prefix granularity for analysis for two reasons. First, /24 is a common block size of BGP routing prefixes based on the observations on BGP routing tables. Based on the block size distribution of BGP prefixes in a recent snapshot of BGP routing table from the RouteView project [13], the /24 blocks account for over 50% of all the total prefixes on the Internet. In addition, multiple /24 prefixes could form larger prefixes by prefix aggregations. For example, two neighboring /24 prefixes could form /23 prefixes, thus the clusters identified in these two /24 prefixes could become separate clusters or be merged together to form a large cluster due to common traffic behavior. Secondly, the prefix-preserving anonymization process makes it impractical to aggregated IP addresses into real BGP prefixes or larger network prefixes. On the other hand, our proposed algorithm could be applied to BGP prefixes if data packets are not anonymized in other datasets. Our analysis is applied to both source and destination prefixes, since the bipartite graphs and one-mode projection graphs in the previous sections could be established for both sides.

To determine an appropriate timescale for analyzing network traffic, we run the proposed algorithms with six different timescales including 10 s, 30 s, 1 min, 2 min, 3 min, and 5 min. Figure 5.7a–c illustrates the number of prefixes (top figure), the average number of hosts per prefix (middle figure), and the average number of clusters per prefix (bottom figure) for these timescales. Apparently, the number of prefixes increases as a result of increasing scale of observations. However, the average number of hosts and clusters per prefix tends to decrease when the timescale increases from 1 min to 2 and more minutes. Our in-depth study reveals that during the longer time windows we tend to observe more single-packet and short-lived flows to a smaller number of random hosts in the same network prefixes due to pervasive scanning activities on the Internet. During time windows with a smaller timescale we mainly observe normal multiple-packet and long-lived flows such as traffic from server farms of popular websites and video streaming services. As a result, the decreasing number of hosts per prefix leads to smaller traffic clusters. The three timescales 10 s, 30 s, and 1 min have the highest numbers of average cluster size. In addition, Fig. 5.8 shows the percentage of hosts in the first cluster over all hosts in the prefix across varying timescales. As the timescale of observation increases, each prefix tends to include additional hosts in the prefix that do not share similar traffic behavior with other hosts. In other words, the increased timescale of observation leads to an increased number of clusters with one or a few hosts. As shown in Fig. 5.8, the timescales 10 s, 30 s, and 1 min have the highest percentages of hosts in the top cluster. Thus,

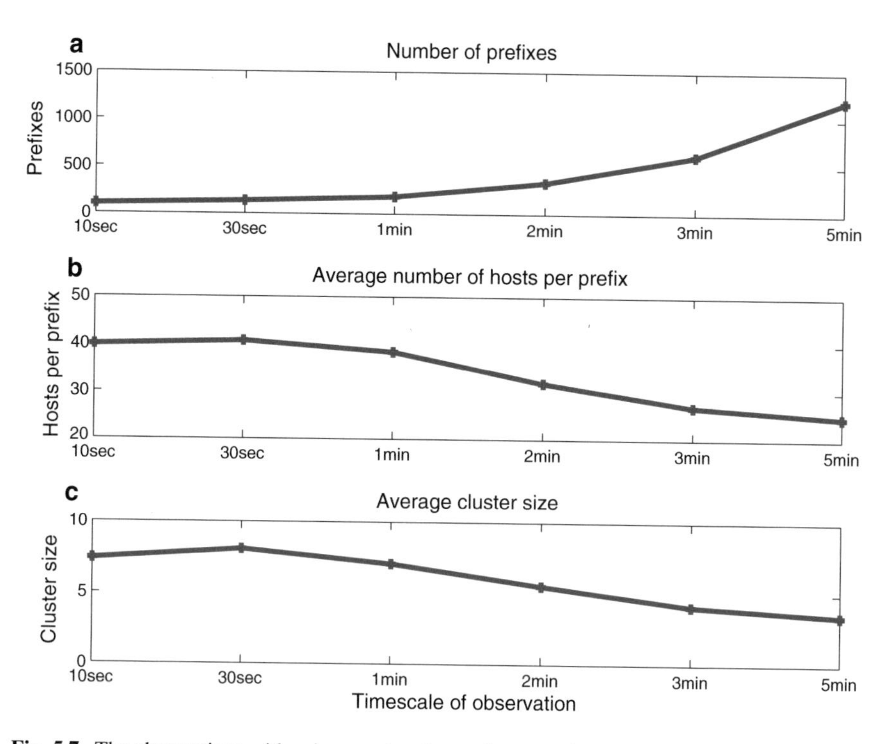

Fig. 5.7 The observations with using varying timescales to analyze 1-hour traffic data: **a** number of prefixes; **b** average number of hosts per prefixes; **c** average cluster size

we consider these three timescales as good candidates for appropriate timescales for traffic analysis.

To evaluate the operational feasibility of the clustering algorithm, we run the clustering process on a commodity Linux server with a 2.93 GHz CPU and 2G memory using the traffic data. Figure 5.9 illustrates the running time of the clustering process in discovering end-host behavior clusters of both source and destination network prefixes for three timescales: 10 s (bottom figure), 30 s (middle figure), and 1 min (top figure). In average, it takes 27.5, 42.9, and 47.8 s to complete the clustering process for both source and destination IP prefixes observed in 10 s, 30 s, and 1 min timescales, respectively. The clustering step is able to keep up with the continuous input of 1-min traffic data, but is unable to keep up with the input of 30 s or 10 s traffic data. Thus, we choose 1 min as the timescale for our further analysis.

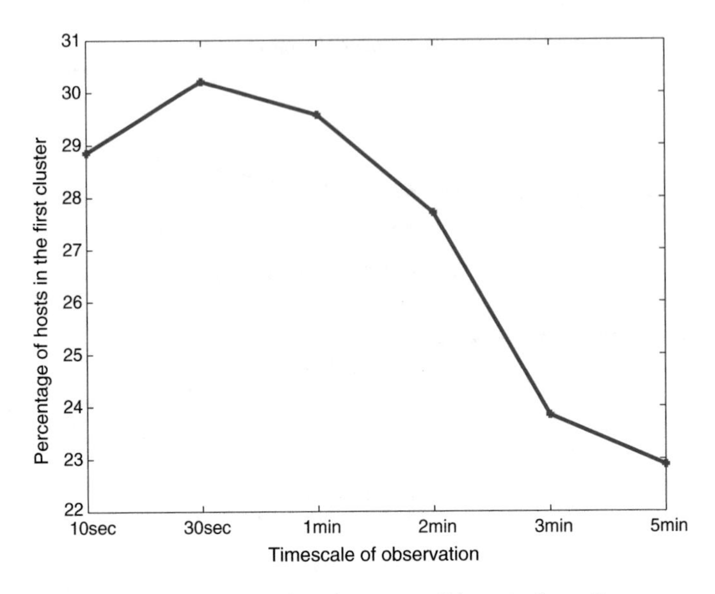

Fig. 5.8 The percentage of hosts in the first cluster over all hosts in the prefix

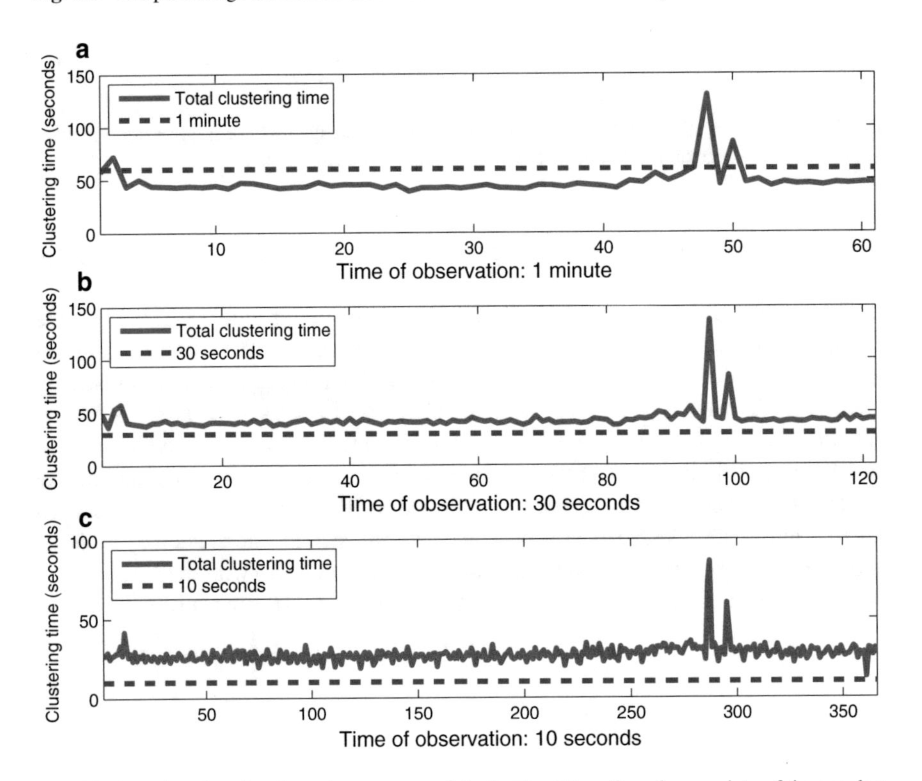

Fig. 5.9 Running time for clustering source and destination IP prefixes for a variety of time scales: **a** 1 min, **b** 30 s, **c** 10 s

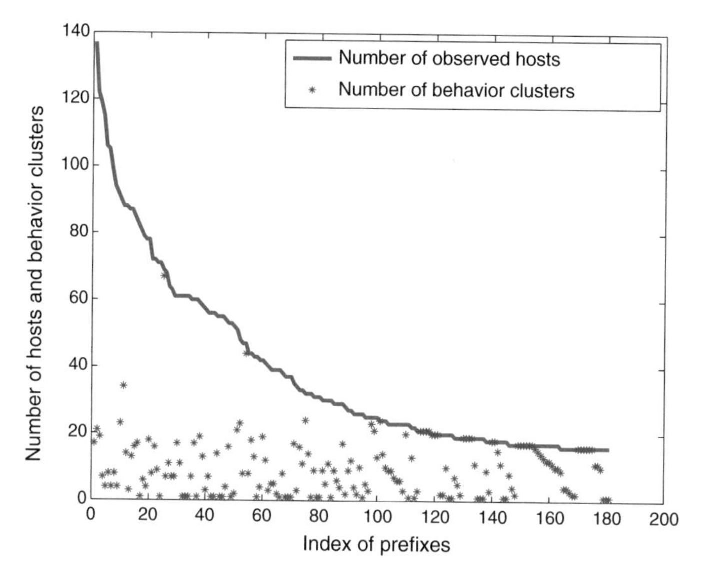

Fig. 5.10 The number of observed hosts and behavior clusters in all the prefixes with at least 16 hosts during 1-min time window

5.5.2 *Distinct Traffic Characteristics of Behavior Clusters*

The network-aware behavior clustering of networked systems shifts traffic analysis from host-level to prefix-level clusters, and increases the granularity of traffic analysis compared with host-level traffic profiling, and thus could successfully reduce the number of behavior profiles for analysis. Figure 5.10 illustrates the size of the prefixes with at least 16 networked systems and the number of their clusters during a 1-min time window. As we can see, the number of clusters is much smaller than the size of prefixes, as each behavior cluster groups many networked systems together due to their common social-behavior patterns. This observation holds for other time windows as well. From Fig. 5.10, it is also interesting to see that there exists little correlation between the number of observed hosts and the number of behavior clusters. The number of behavior clusters for an IP prefix largely depends on the similarity of the social behavior patterns among the observed hosts, rather than the count of observed hosts. For example, the IP prefixes of data center networks that include hundreds of servers tend to have less diverse behavior, while the IP prefixes of residential Internet service providers could have more diverse behavior since the hosts in residential networks could have very different communication patterns.

After obtaining separate behavior clusters, the next question we ask is *do networked system behavior clusters indeed exhibit distinct traffic characteristics?* Towards answering this question, we study the distributions of traffic features in each of behavior clusters, and then compare them with the aggregated traffic of the prefixes. We use relative uncertainty (RU), which is an information-theoretic mea-

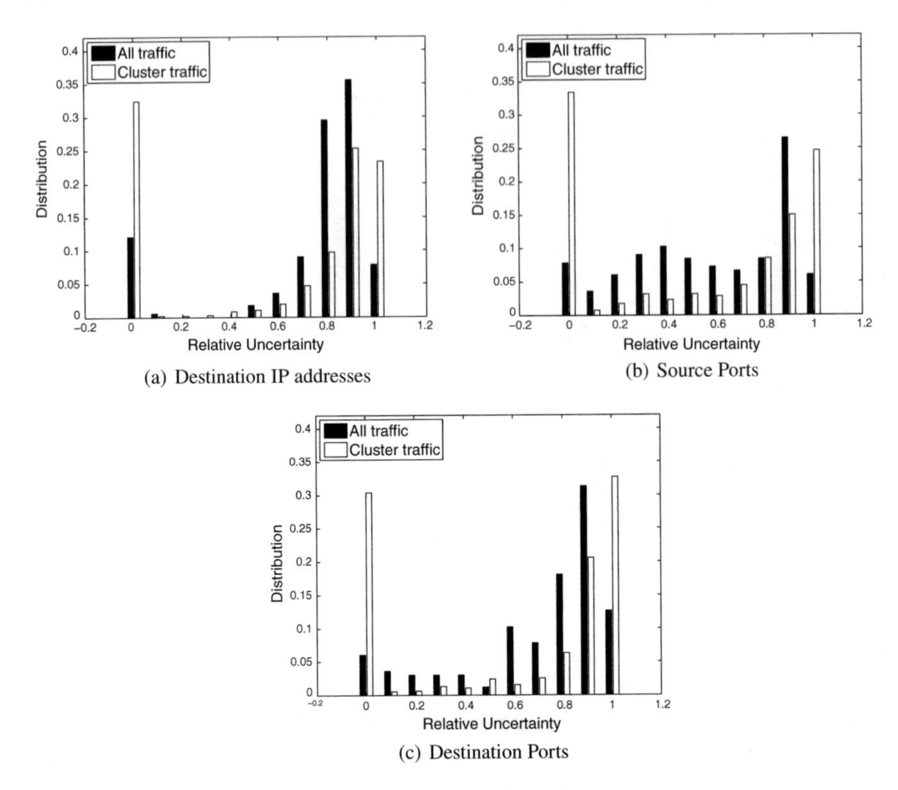

(a) Destination IP addresses

(b) Source Ports

(c) Destination Ports

Fig. 5.11 Histograms of relative uncertainty distributions for behavior clusters and the aggregated traffic

sure explained in Sect. 2.3.1, to analyze the traffic features in individual clusters and the aggregated traffic.

Our results show that behavior clusters separate different traffic patterns of the same prefixes for improved understanding and interpretation. Figure 5.11 shows the distribution of relative uncertainty on destination IP addresses, source ports, and destination ports, respectively, for all the source prefixes and their behavior clusters during a 1-min time window. Compared with relative uncertainty values for network prefixes, the behavior clusters have much larger percentages of relative uncertainty values on all of these features being 0 and 1 or approximately being 0 and 1, which reveal concentrated patterns on a few ports and IP addresses, or random patterns on ports and addresses. This result shows that the clustering algorithm extracts behavior clusters with distinct traffic characteristics from the aggregated traffic in the network prefixes, thus significantly improving the understanding of the traffic patterns with detailed and meaningful interpretations.

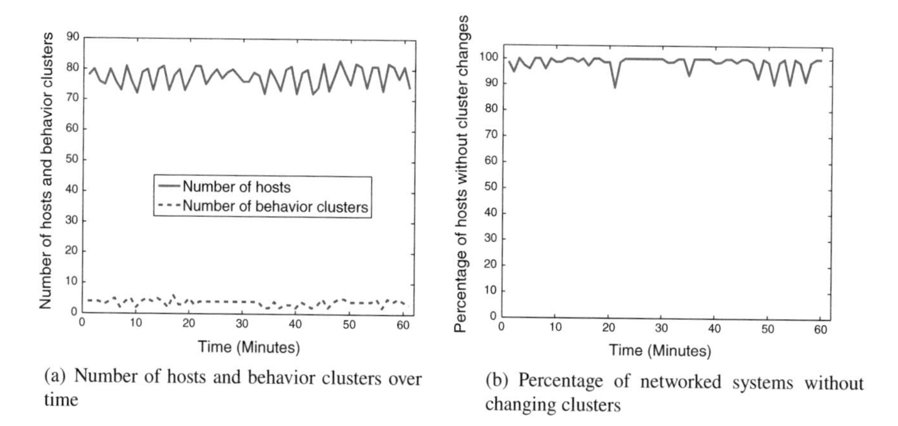

(a) Number of hosts and behavior clusters over time

(b) Percentage of networked systems without changing clusters

Fig. 5.12 Temporal stability of behavior clusters in a network prefix

5.5.2.1 Temporal Stability of Behavior Clusters

The second question on the characteristics of end-host behavior clusters we ask is *are the clusters stable over time?* In other words, *do networked systems in the same prefixes change clusters over time?* To address this question, we study the temporal stability of behavior clusters and the dynamics of cluster changes for networked systems over time. Figure 5.12a illustrates the high temporal stability of behavior clusters for one IP prefix during the 1-hour time window. As shown by the top line in Fig. 5.12a, the number of networked systems in the prefix fluctuates slightly over time, since some hosts do not continuously send or receive traffic. More importantly, the number of behavior clusters, illustrated by the bottom line in Fig. 5.12a, also exhibits slight fluctuations over time. Similar observations hold for other prefixes.

In addition, we find the majority of networked systems stay in the same behavior cluster over time. Figure 5.12b shows the high percentage of networked systems in the network prefix in Fig. 5.12a without changing clusters over consecutive 1-min time windows. In average, 71.8% of all the networked systems in the traffic traces do not change clusters during the 1-hour time period. Our experiments with varying timescales also show similar observations hold for other timescales as well. For example, Fig. 5.13a, b illustrate the high percentages of networked systems do not change clusters over continuous 30-s and 2-min time windows, respectively. These observations confirm that network-aware behavior clustering separates networked systems of network prefixes into distinct and stable behavior clusters.

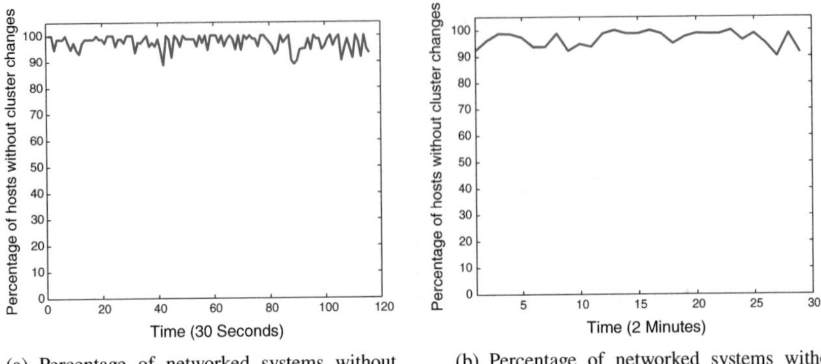

(a) Percentage of networked systems without changing clusters (30 seconds)

(b) Percentage of networked systems without changing clusters (2 minutes)

Fig. 5.13 Temporal stability of behavior clusters in a network prefix for different timescales

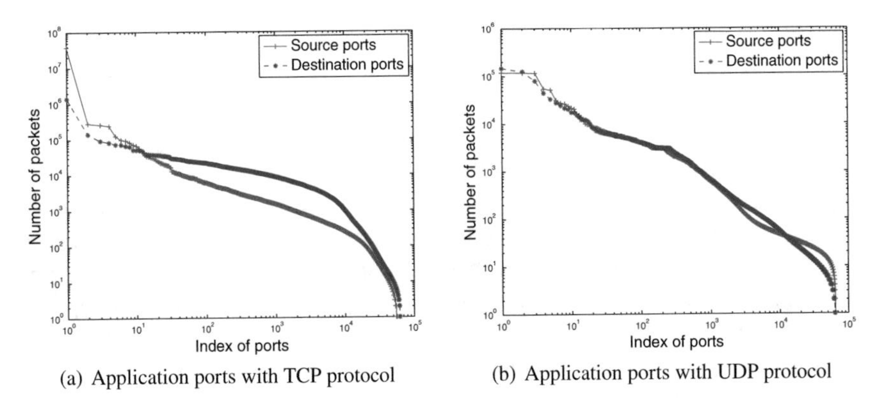

(a) Application ports with TCP protocol

(b) Application ports with UDP protocol

Fig. 5.14 Distribution of network traffic for Internet applications observed from Internet backbone links

5.5.3 *Exploring Similarity of Internet Applications*

Figure 5.14a, b illustrates the distribution of IP packets for Internet application traffic observed from one backbone link during 1-min time window for TCP and UDP ports, respectively. It is interesting to observe that a large number of application ports, regardless transport protocols (TCP or UDP) and traffic directions (source ports or destination ports), carry non-trivial data traffic. For example, there are over 2550 TCP destination ports with more than 5000 IP packets on the link during the 1-min time window. In other words, the traditional top N approaches of focusing on a few top ports with the largest amount of traffic is not sufficient, since it is also very important to study the other applications with significant volumes of IP traffic.

Building source and destination behavior graphs for each application port in our proposed method provides an opportunity to understand the social behavior of source

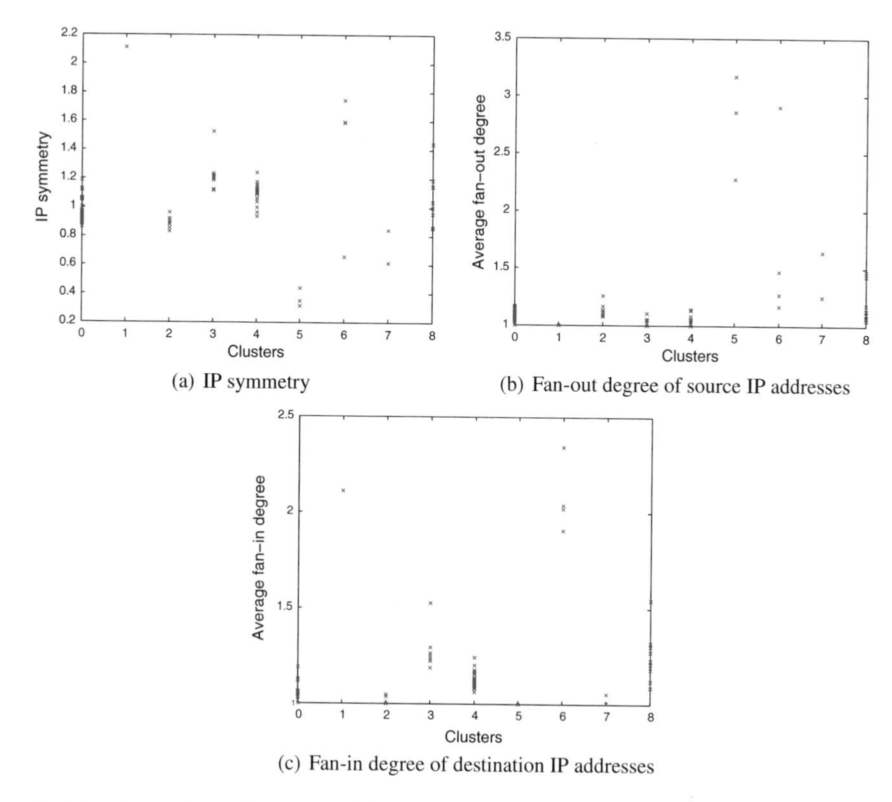

(a) IP symmetry

(b) Fan-out degree of source IP addresses

(c) Fan-in degree of destination IP addresses

Fig. 5.15 Distinctive traffic characteristics of application clusters

and destination hosts engaging in the same applications. In addition, grouping these applications based on clustering coefficient of source and destination behavior graphs into distinct clusters helps understand unknown applications that share similar patterns with well-known applications.

To evaluate the quality of the clustering results, we study traffic characteristics of application clusters and compare the similarity in traffic characteristics among application ports in the same clusters as well as the dissimilarity among ports in different clusters. Our experiment results show that the application clusters indeed exhibit distinctive traffic characteristics. Specifically, for each application port p, we study IP symmetry $ipsym_p$, fan-out degree of source hosts $fanout_p$, and fan-in degree of destination hosts $fanin_p$. The IP symmetry $ipsym_p$ is given by the ratio between unique source hosts and unique destination hosts engaging in the application port p. The fan-out degree for the application is the average fan-out degree of all source hosts involving in the application, while the fan-in degree is the average fan-in degree of all destination hosts.

Figure 5.15a–c illustrates the distinctive characteristics in IP symmetry, fan-out degrees of source hosts, and fan-in degree of destination hosts for application clusters during one time window, respectively. The similar observations hold for other time windows as well. This observation confirms that our proposed method of behavioral graph analysis on Internet applications is indeed able to discover distinct clusters of application ports that not only exhibit similar clustering coefficient in their source and destination behavior graphs, but also share similar traffic characteristics in IP symmetry, fan-out and fan-in degrees.

5.6 Summary

This chapter studies graphical models for characterizing traffic patterns of networked systems and Internet applications, specifically explores bipartite graphs and one-mode projection graphs to analyze social behavior of networked systems within the same network prefixes or engaging in the same Internet applications. By applying clustering algorithms on the similarity matrices of one-mode projection graphs, we find the clustered behavior of end hosts in the same network prefixes. Through clustering coefficient and other graph properties, we also find interesting similarity of social behavior among different Internet applications and discover distinctive application behavior clusters that group applications with similar social behavior. The practical benefits and applications of exploring behavior similarity and discovering host and application behavior clusters include profiling network behaviors, discovering emerging network applications, and detecting anomalous traffic patterns. In addition, the idea of applying bipartite graphs and one-mode projections has also been successfully extended to analyze other networks such as online social networks [14].

References

1. H. Jiang, Z. Ge, S. Jin, J. Wang, Network prefix-level traffic profiling: characterizing, modeling, and evaluation. Comput. Netw. (2010)
2. K. Xu, Z.-L. Zhang, S. Bhattacharyya, Profiling internet backbone traffic: behavior models and applications, in *Proceedings of ACM SIGCOMM* (2005)
3. S. Wei, J. Mirkovic, E. Kissel, Profiling and clustering internet hosts, in *Proceedings of the International Conference on Data Mining* (2006)
4. Y. Jin, E. Sharafuddin, Z.-L. Zhang, Unveiling core network-wide communication patterns through application traffic activity graph decomposition, in *Proceedings of ACM SIGMETRICS* (2009)
5. J.-L. Guillaume, M. Latapy, Bipartite graphs as models of complex networks. Phys. A: Stat. Theor. Phys. **371**(2), 795–813 (2006)
6. A. Ng, M. Jordan, Y. Weiss, On spectral clustering: analysis and an algorithm, in *Proceedings of Neural Information Processing Systems (NIPS) Conference* (2001)
7. J. Ramasco, S. Dorogovtsev, P. Romualdo, Self-organization of collaboration networks. Phys. Rev. **70**(3) (2004)
8. M. Kaiser, Mean clustering coefficients: the role of isolated nodes and leafs on clustering measures for small-world networks. New J. Phys. **10** (2008)

9. P.-N. Tan, M. Steinbach, V. Kumar, *Introduction to Data Mining* (Addison-Wesley, New York, 2006)
10. Cooperative Association for Internet Data Analysis (CAIDA): Internet Traces, http://www.caida.org/data/passive/passive_2009_dataset.xml
11. J. Fan, J. Xu, M. Ammar, S. Moon, Prefix-preserving IP address anonymization: measurement-based security evaluation and a new cryptography-based scheme. Comput. Netw. **46**(2), 253–272 (2004)
12. K. Xu, Z.-L. Zhang, S. Bhattacharyya, Internet traffic behavior profiling for network security monitoring. IEEE/ACM Trans. Netw. **16**, 1241–1252 (2008)
13. U. Oregon, Routeviews archive project, http://archive.routeviews.org/
14. F. Wang, K. Xu, H. Wang, Discovering shared interests in online social networks, in *Proceedings of IEEE ICDCS Workshop on Peer-to-Peer Computing and Online Social Networking (HOTPOST). Macao, China* (2012)

Chapter 6
Real-Time Network Behavior Analysis

Abstract Recent years have seen significant progress in real-time, continuous network traffic monitoring and measurement systems on the Internet. However, real-time traffic summaries reported by many such systems focus mostly on volume-based heavy hitters, which are not sufficient for finding interesting or anomalous behavior patterns. This chapter discusses the feasibility of building a real-time network behavior analysis system that analyzes vast amounts of traffic data in IP networks and reports comprehensive behavior patterns of networked systems and Internet applications. This chapter first discusses the importance and challenges of building real-time network behavior analysis systems. Subsequently, this chapter presents the real-time network behavior analysis system and discusses its functional modules as well as the interfaces with continuous monitoring systems and an event analysis engine, and discusses the performance benchmarking and stress test of the real-time system using a variety of packet-level traces from Internet backbone links, and synthetic traces that mix various attacks into real backbone packet traces. Finally, this chapter introduces and evaluates sampling-based filtering algorithms to enhance the robustness of the network behavior analysis system against sudden traffic surges.

6.1 Real-Time Network Measurement and Monitoring

Recent years have seen significant progress in real-time, continuous network traffic monitoring and measurement systems in IP networks [1, 2]. However, *real-time* traffic summaries reported by many such systems focus mostly on volume-based heavy hitters (e.g., top N ports or IP addresses that send or receive most traffic) or aggregated metrics of interest (total packets, bytes, flows, etc.) [3], which are not sufficient for finding interesting or anomalous behavior patterns. In this chapter, we explore the feasibility of building a real-time network behavior analysis system that characterizes and models vast amount of traffic data in an IP backbone network and reports *comprehensive behavior patterns* of networked systems and Internet applications.

Towards this end, we answer a specific question in this chapter: is it feasible to build a *robust* real-time network behavior analysis system that is capable of con-

K. Xu, *Network Behavior Analysis*,
https://doi.org/10.1007/978-981-16-8325-1_6

tinuously extracting and analyzing "interesting" and "significant" traffic patterns on high-speed (OC48 or higher speed) Internet links, even in the face of sudden surge in traffic (e.g., when the network is under a denial-of-service attack)? We address this question in the context of the network behavior analysis framework we have developed for IP networks. The behavior and structural models in the framework employ a combination of data-mining and information-theoretic techniques to build comprehensive behavior profiles of Internet backbone traffic in terms of communication patterns of networked systems and Internet applications. It consists of three key steps: significant cluster extraction, automatic behavior classification, and structural modeling for in-depth interpretive analysis. This three-step network behavior analysis framework extracts networked systems and Internet applications that generate significant traffic, classifies them into different *behavior classes* that provide a general separation of various *common* "normal" (e.g., web server and service traffic) and "abnormal" (e.g., scanning, worm, or other exploit traffic) traffic as well as *rare* and anomalous traffic behavior patterns. The framework has been extensively validated *offline* using packet traces collected from a variety of backbone links in an IP backbone network.

To demonstrate the operational feasibility of performing *online* network behavior analysis on high-speed Internet backbone links, we build a prototype system [4, 5] using general-purpose commodity PCs and integrate it with an existing real-time traffic monitoring and collection system operating in an Internet backbone network. The real-time traffic monitoring and collection system captures packets on a high-speed link (from OC12 to OC192) and converts them into 5-tuple flows (based on source IP, destination IP, source port, destination port, protocol fields), which are then continuously fed to the real-time network behavior analysis system we built. The large volume of traffic flows observed from these links creates great challenges for the network behavior analysis system to process them *quickly* on commodity PCs with *limited memory* capacity. We incorporate several optimization features in our implementation such as efficient data structures for storing and processing cluster information to address these challenges.

6.2 Real-Time System for Network Behavior Analysis

In this section, we first describe the design guidelines for our network behavior analysis system and then present the overall architecture, functional modules, and some key implementation details.

6.2.1 Design Guidelines

The design and implementation of our network behavior analysis system follow four key principles: high scalability, strong robustness, flexible modularity, and great usability:

- **High Scalability:** The network behavior analysis system is targeted at high-speed (1 Gbps or more) backbone links and hence must scale to the traffic load offered by such links. Specifically, if the system has to continuously build behavior profiles of significant clusters once every time interval T (e.g., $T = 5\,\text{min}$), then it has to take less than time T to process all the flow records aggregated in every time interval T. And this has to be accomplished on a commodity PC platform.
- **Strong Robustness:** The network behavior analysis system should be robust to anomalous traffic patterns such as those caused by denial-of-service attacks, flash crowds, worm outbreaks, etc. These traffic patterns can place a heavy demand on system resources. At the same time, it is vital for the network behavior analysis system to be functioning during such events since it will generate data for effective response and forensic analysis. Therefore, the system must adapt gracefully to these situations and achieve a suitable balance between the accuracy of behavioral and structural models and resource utilization.
- **Flexible Modularity:** The system should be designed in a modular fashion with each module encapsulating a specific function or step in the network behavior analysis framework. Information exchange between modules should be clearly specified. In addition, the system should be designed to accept input from any packet or flow monitoring system that exports a continuous stream of flow records. However, the flow record export format has to be known to the system.
- **Great Usability:** The network behavior analysis system should be easy to configure and customize so that a network operator can focus on specific events of interest and obtain varying levels of information about these events. At the same time, it should expose minimal details about the methodology to an average user. Finally, it should generate meaningful and easy-to-interpret event reports, instead of streams of statistics.

These design considerations form a guideline of our system design and drive each stage of our system implementation. In the rest of the section, we will discuss the overall architecture of the real-time network behavior analysis system, its functional modules, and key implementation details that achieve design goals.

6.2.2 System Architecture

Figure 6.1 depicts the architecture of the network behavior analysis system that is integrated with an "always-on" monitoring system and an event analysis engine. The flow-level information used by the network behavior analysis system are generated

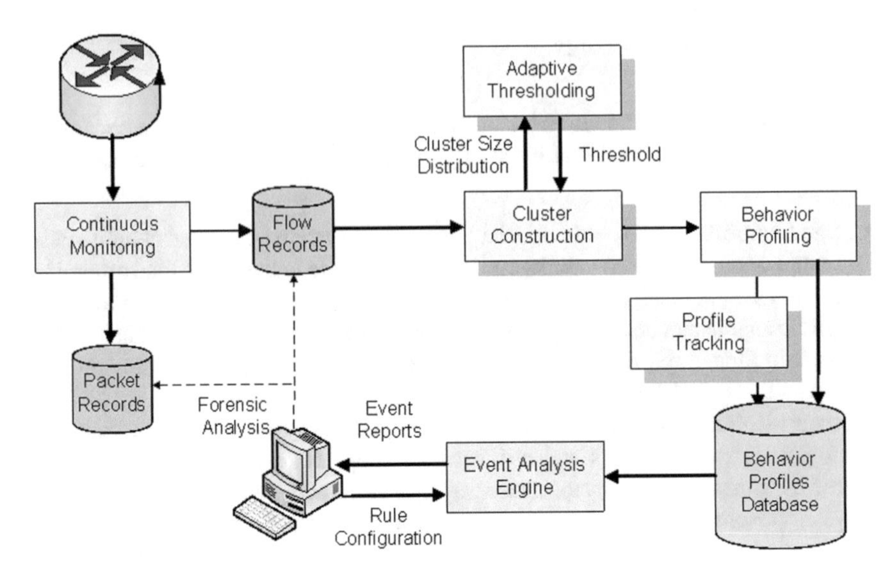

Fig. 6.1 System architecture for real-time network behavior analysis

from continuous packet or flow monitoring systems that capture packet headers on a high-speed Internet link via an optical splitter and a packet capturing device, i.e., DAG card. The monitoring system aggregates packets into 5-tuple flows and exports the flow records for a given time interval into disk files. In general, the network behavior analysis system obtains flow records through three ways: (i) shared disk access, (ii) file transfer over socket, and (iii) flow transfer over a streaming socket. The option in practice will depend on the locations of the profiling and monitoring systems. The first way works when both systems run on the same machine, while the last two can be applied if they are located in different machines.

In order to improve the efficiency of the network behavior analysis system, we use distinct process threads to carry out multiple task in parallel. Specifically, one thread continuously reads flow records in the current time interval T_i from the monitoring systems, while another thread profiles flow records that are complete for the previous time interval T_{i-1}.

The event analysis engine analyzes a *behavior profile database*, which includes current and historical behavior profiles of end hosts and network applications reported by the *behavior profiling* and *profile tracking* modules in the network behavior analysis system.

The real-time network behavior analysis system consists of four functional modules (shadowed boxes), namely, "cluster construction", "adaptive thresholding", "behavior profiling", and "profile tracking". These four modules are responsible for constructing traffic clusters for behavioral entities observed from traffic data, identifying and extracting significant networked systems and Internet applications based on adaptive thresholds, developing behavioral and structure models of these systems and applications, and tracking the dynamics of network behaviors over time, respectively. Their main functions are briefly summarized below:

- **Cluster Construction:** This module has two initialization tasks. First it starts to load a flow table (FTable) in a time interval T into memory once the network behavior analysis system receives a signal indicating FTable is ready. The second task is to group flows in FTable associated with the same feature values (i.e., cluster keys) into *clusters*.
- **Adaptive Thresholding:** This module analyzes the distribution of *flow counts* in the four feature dimensions and computes a threshold for extracting significant clusters along each dimension.
- **Behavior Profiling:** This module implements a combination of behavior classification and structural modeling that builds behavior profiles in terms of communication patterns of significant end hosts and applications.
- **Profile Tracking:** This module examines all behavior profiles built from the network behavior analysis system from various aspects to find interesting and suspicious network events.

6.2.3 Key Implementation Details

6.2.3.1 Data Structures

High-speed backbone links typically carry a large amount of traffic flows. Efficiently storing and searching these flows is critical for the *scalability* of our real-time network behavior analysis system. We design two efficient data structures, namely, FTable and CTable for efficient storage and fast lookups during cluster extraction and behavior modeling.

Figure 6.2 illustrates the data structure of FTable and CTable with an example. FTable, an array data structure, provides an index of 5-tuple flows through a commonly used hash function,

$$FH = srcip^\wedge dstip^\wedge srcport^\wedge dstport^\wedge proto\%(FTableEntries - 1), \quad (6.1)$$

where $FTableEntries$ denotes the maximum entries of FTable. For example, in Fig. 6.2, *flow 1* is mapped to the entry 181 in FTable, while *flow 2* is mapped to the entry 1. In case of hashing collision, i.e., two or more flows mapping to the same table entry, we use a linked list to manage them. In our experiments, the (average) collision rate of this flow hash function is below 5% with $FTableEntries = 2^{20}$. While constructing clusters, the naive approach would be to make four copies of 5-tuple flows, and then group each flow into four clusters along each dimension. However, this method dramatically increases the memory cost of the system since the flow table typically has hundreds or millions of flows in each time interval. Instead of duplicating flows, which is expensive, we add four flow pointers (i.e., next

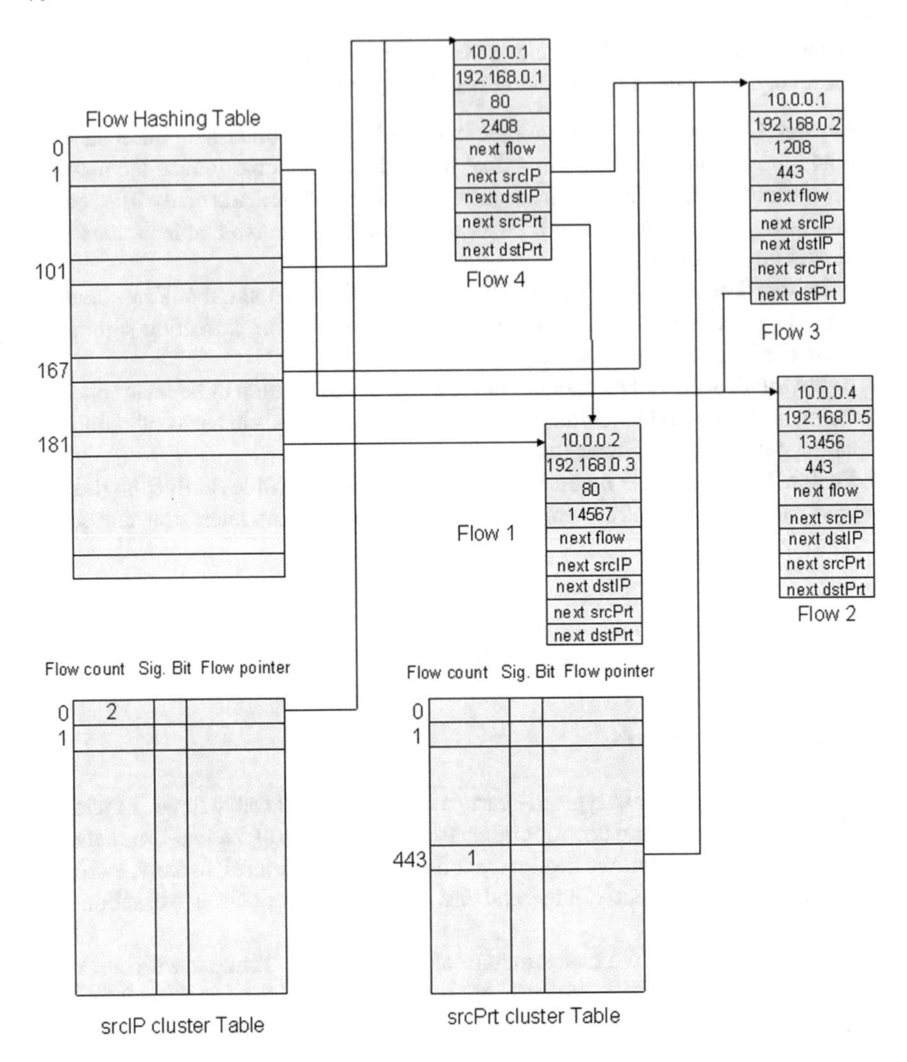

Fig. 6.2 Data structure of flow table and cluster table

`srcIP`, `next dstIP`, `next srcPrt`, and `next dstPrt`) in each flow. Each flow pointer will link the flows sharing the same feature value in the given dimension. For example, the `next srcIP` pointer of *flow 4* links to `flow 3` since they share the same `srcIP` *10.0.0.1*. Similarly, the `next srcPrt` pointer of *flow 4* links to `flow 1` since they share the same `srcPrt` *80*. However, the question is how to quickly find the "old" flows of the same clusters when adding a new flow in the flow table.

To address this problem, we create another data structure, `CTable`, which links the first flow of each cluster in `FTable`. Since there are four types of clusters, we create four instances of `CTable` for managing clusters along four dimensions.

Table 6.1 Notations used in the complexity analysis

Notation	Definition
F	set of 5-tuple flows in a time interval
i	dimension id (0/1/2/3 = `srcIP/dstIP/srcPort/dstPort`)
C_i	set of clusters in dimension i
S_i	set of significant clusters in dimension i
c_i	a cluster in dimension i
s_i	a significant cluster in dimension i
r_f	size of a flow record
r_v	size of the volume information of a cluster
r_b	size of behavior information of a sig. cluster
r_s	size of dominant states of a significant cluster

Considering `srcPrt` and `dstPrt` dimensions with 65536 possible clusters (ports), we use an array with a size of 65536 to manage the clusters for each of these two dimensions. The index of the array for each port is the same as the port number. For `srcIP` and `dstIP` dimensions, we use a simple hash function that performs a bitwise exclusive OR (XOR) operation on the first 16 bits and the last 16 bits of `IP` address to map each `srcIP` or `dstIP` into its `CTable` entry. When adding a new flow, e.g., `flow 3` in Fig. 6.2, in the given `dstPrt`, we first locate the first flow (`flow 2`) of the cluster `dstPrt` 443 and make the `next dstPrt` pointer of `flow 3` to `flow 2`. Finally, the first flow of the cluster `dstPrt` 443 is updated to `flow 3`. This process is similar for the cluster `srcPrt` 1208, as well as the the clusters `srcIP` *10.0.0.1* and `dstIP` *192.168.0.2*.

In addition to pointing to the first flow in each cluster, each `CTable` entry also includes flow count for the cluster and significant bit for marking significant clusters. The former maintains flow counts for cluster keys, and the flow count distribution will determine the adaptive threshold for extracting significant clusters.

6.2.3.2 Space and Time Complexity of Modules

The space and time complexity of modules essentially determines the CPU and memory cost of the network behavior analysis system. Thus, we quantify the complexity of each module in the system. For convenience, Table 6.1 shows the definitions of the notations that will be used in the complexity analysis.

The time complexity of cluster construction is

$$O(|F| + \sum_{i=0}^{3} |C_i|) \tag{6.2}$$

for `FTable` and `CTable` constructions. Similarly, the space complexity is

$$O(|F| * s_{fr} + \sum_{i=0}^{3}(|C_i| * r_v)). \tag{6.3}$$

The time complexity of adaptive thresholding is

$$\sum_{i=0}^{3}(|C_i| * e_i). \tag{6.4}$$

This module does not allocate additional memory, since its operations are mainly on the existing `CTable`. Thus, the space complexity is zero.

The time complexity of behavior profiling is

$$O\left(\sum_{i=0}^{3}\sum_{j=0}^{|S_i|}|s_j|\right), \tag{6.5}$$

while the space complexity is

$$O\left(\sum_{i=0}^{3}[|S_i| * (r_b + r_s)]\right). \tag{6.6}$$

The output of this step is the behavior profiles of significant clusters, which are recorded into a database along with the timestamp for further analysis.

Due to a small number of significant clusters extracted, the computation complexity of profile tracking is often less than the others in two or three orders of magnitude, so for simplicity we will not consider its time and space requirement.

6.2.3.3 Parallelization of Input and Profiling

In order to improve the efficiency of the network behavior analysis system, we use *thread* mechanisms for parallelizing tasks in multiple modules, such as continuously importing flow records in the current time interval T_i, and profiling flow records that are complete for the previous time interval T_{i-1}. Clearly, the parallelization could reduce the time cost of the network behavior analysis system. The disadvantage of doing so is that we have to maintain two set of `FTable` and `CTable` for two consecutive time intervals.

Table 6.2 Total CPU and memory cost of the real-time network behavior analysis system on 5-min flow traces

Link	Util.	CPU time (sec)			Memory (MB)		
		min	avg	max	min	avg	max
L_1	207 Mbps	25	46	65	82	96	183
L_2	86 Mbps	7	11	16	46	56	71
L_3	78 Mbps	7	12	82	45	68	842

6.2.3.4 Event Analysis Engine

To discover interesting or suspicious network events, we build an event analysis engine with three aspects: (i) temporal behavior analysis, (ii) feature dimension correlation, and (iii) event configurations. The objective of temporal behavior analysis is to characterize temporal properties of behavior classes as well as individual clusters from the behavior profile database that records behavior profiles built from the network behavior analysis system. Prior work in [6, 7] has demonstrated that temporal properties could help distinguish and classify behavior classes. Feature dimension correlation attempts to find the correlation between clusters from various dimensions to detect emerging exploit and worm activities [8–10] that often trigger new clusters from `srcIP`, `dstIP`, and `dstPrt` dimensions.

We develop a simple *event configuration* language that enables network operators or security analysts to extract information on events of interest from behavior profiles for network management or troubleshooting. To express the policy, we use four distinct fields: Dimension, Event Type, Filter, and Description. The options of these fields include the following:

- Dimension = srcIP | dstIP | srcPrt | dstPrt | all
- Event Type = rare | deviant | exploit | unusual service ports | all
- Filter = high frequency | high intensity | matching selected ports | ......
- Description = full | summary

For example, if a network operator wants to monitor *rare* behavior of `srcIP` end hosts, she could use the rule srcIP (Dimension) AND rare (Event Type) AND all (Filter) AND full (Description), which expresses the policy of reporting *full* profiles of *all* `srcIP` clusters with *rare* behavior. Similarly, we could construct other filter rules using the combinations of all available options.

In the next section, we will demonstrate the operational feasibility of this system by performing extensive benchmarking of CPU and memory costs using packet-level traces from OC48 backbone links. To evaluate the robustness of the system, we also test the system against anomalous traffic patterns under emulated denial-of-service attacks or worm outbreaks.

6.3 Performance Evaluation

After designing and implementing this real-time network behavior analysis system, we perform extensive benchmarking of CPU and memory costs using packet-level traces from Internet backbone links to identify the potential challenges and resource bottlenecks. We find that CPU and memory costs increase linearly with number of flows seen in a given time interval. Nevertheless, resources on a commodity PC are sufficient to continuously process flow records and build behavior profiles for high-speed links in operational networks. For example, on a dual 1.5 GHz PC with 2048 MB of memory, building behavior profiles once every 5 min for an 2.5 Gbps link loaded at 209 Mbps *typically* takes 45 s of CPU time and 96 MB of memory.

In this section, we first conduct performance benchmarking of CPU and memory cost of the network behavior analysis system using a variety of packet traces from OC48 backbone links. Subsequently, we evaluate the performance bottlenecks of the system under anomalous traffic patterns such as those caused by denial-of-service attacks and worm outbreaks.

6.3.1 Benchmarking

We measure CPU usage of the behavioral profiling process by using a system call, namely, *getrusage()*, which queries actual system and user CPU time of the process. The system call returns with the resource utilization including *ru_utime* and *ru_stime*, which represent the user and system time used by the process, respectively. The sum of these two times indicates the total CPU time that the profiling process uses. Let T denote the total CPU time, and T_l, T_a, and T_p denote the CPU usage for the modules of cluster construction, adaptive thresholding, and behavior profiling, respectively. Then we have

$$T = T_l + T_a + T_p. \tag{6.7}$$

Similarly, we collect memory usage with another system call, *mallinfo()*, which collects information of the dynamic memory allocation. Let M denote the total memory usage, and M_l, M_a, and M_p denote the memory usage in three key modules. Then we have

$$M = M_l + M_a + M_b. \tag{6.8}$$

In order to track the CPU and memory usages of each module, we use these two system calls before and after the module. The difference of the output becomes the actual CPU and memory consumption of each module. Next, we show the CPU time and memory cost of the network behavior analysis system on three OC48 links during a continuous 18-hour period with an average link utilization of 209 Mbps, 86 Mbps, and 78 Mbps. For convenience, let L_1, L_2, and L_3 denote these three links, respectively.

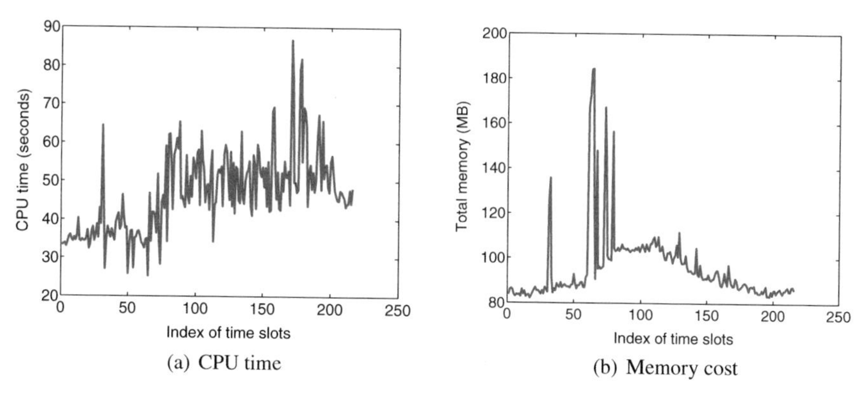

Fig. 6.3 CPU and memory cost of the real-time network behavior analysis system on flow records in 5-min time interval collected in L_1 for 18 consecutive hours

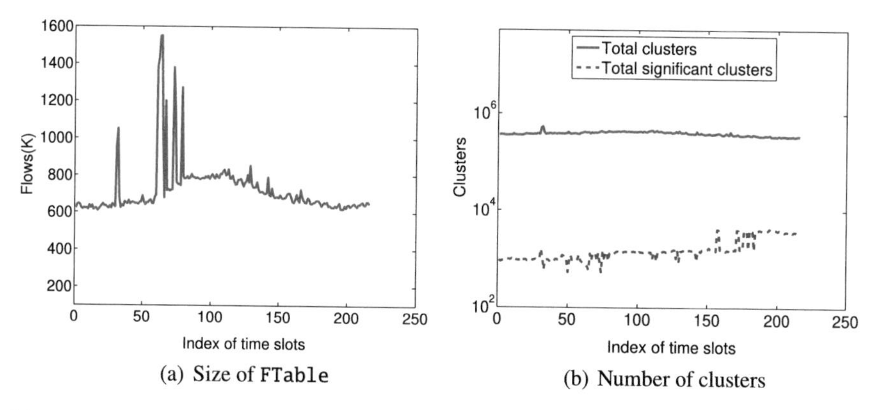

Fig. 6.4 Input of flow traces in 5-min time interval collected in L_1 for 18 consecutive hours

Table 6.2 shows a summary of CPU time and memory cost of the network behavior analysis system on L_1 to L_3 for 18 consecutive hours. It is not surprising to see that the average CPU and memory costs for L_1 are larger than the other two links due to a higher link utilization. Figure 6.3 shows the CPU and memory cost of the system on all 5-min intervals for L_1 (the link with the highest utilization). For the majority of time intervals, the network behavior analysis system requires less than 60 s (1 min) of CPU time and 150 MB of memory using the flow records in 5-min time intervals for L_1.

Figure 6.4a further illustrates the number of flow records over time that ranges from 600K to 1.6M, while Fig. 6.4b shows the number of all clusters as well as the extracted significant clusters. It is very interesting to observe the similar patterns in the plot of memory cost (Fig. 6.3b) and that of the flow count over time (Fig. 6.4a). This observation leads us to analyze the correlation between these two measurements. By examining the breakdown of the memory cost, we find that M_l in the cluster construction module accounts for over 98% of the total memory consumptions. Recall

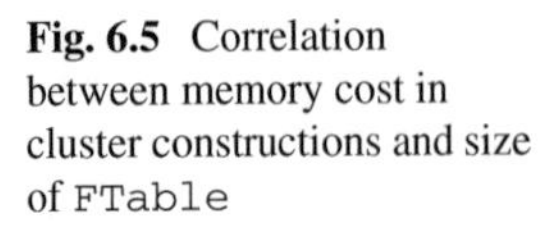
Fig. 6.5 Correlation between memory cost in cluster constructions and size of `FTable`

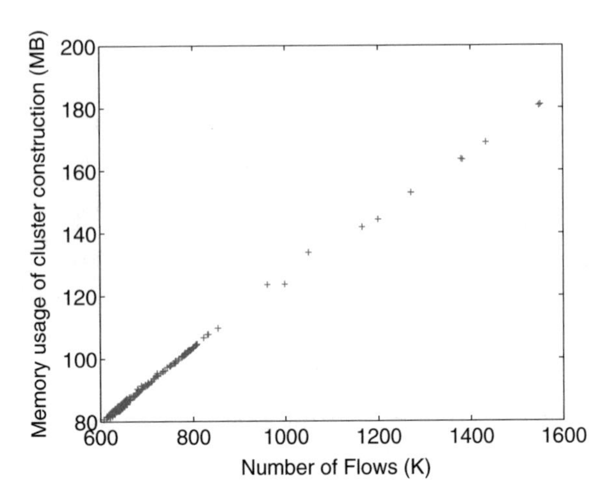

that the space complexity of this module is larger than the others by two or three orders of magnitude, and dominated by the size of flow table $|F|$. The scatter plot of $|F|$ vs. M_l in Fig. 6.5 reflects the linear relationship between them. Therefore, this strong correlation suggests that the memory cost of the network behavior analysis system is mainly determined by the number of flow records collected by the monitoring system in a given time interval.

Figure 6.6a shows a breakdown in CPU usage of the various modules in the network behavior analysis system, and suggests that cluster construction and behavior profiling account for a large fraction of CPU time. Similar to the space complexity, the time complexity in cluster construction is also determined by $|F|$. The linear relationship demonstrated by the scatter plot of $|F|$ versus T_l in Fig. 6.6b confirms this complexity analysis. Figure 6.6c shows the scatter plot of the number of significant clusters versus CPU time in behavior profiling. Overall, we observe an approximately linear relationship between them. This suggests that the CPU cost in behavior profiling is largely determined by the number of significant clusters whose behavior patterns are being analyzed. Recent work [11, 12] has developed efficient algorithms computing information-theoretic functions such as entropy and relative uncertainty on data streams. Thus, the time consumptions of our network behavior analysis system could potentially be improved if these algorithms are incorporated into the system.

To understand how performance is affected by time granularity, we also evaluate the system on L_1 using 1-min, 2-min, 10-min, and 15-min flow traces. The results are shown in Table 6.3. In general, the CPU time and memory cost increase as the length of the time interval. On the other hand, the CPU time of the network behavior analysis system is always less than the time interval T. In addition, the average memory cost for 5-min, 10-min, and 15-min are 96 MB, 151 MB, and 218 MB, respectively, which are within the affordable range on commodity PCs. These results clearly suggest that

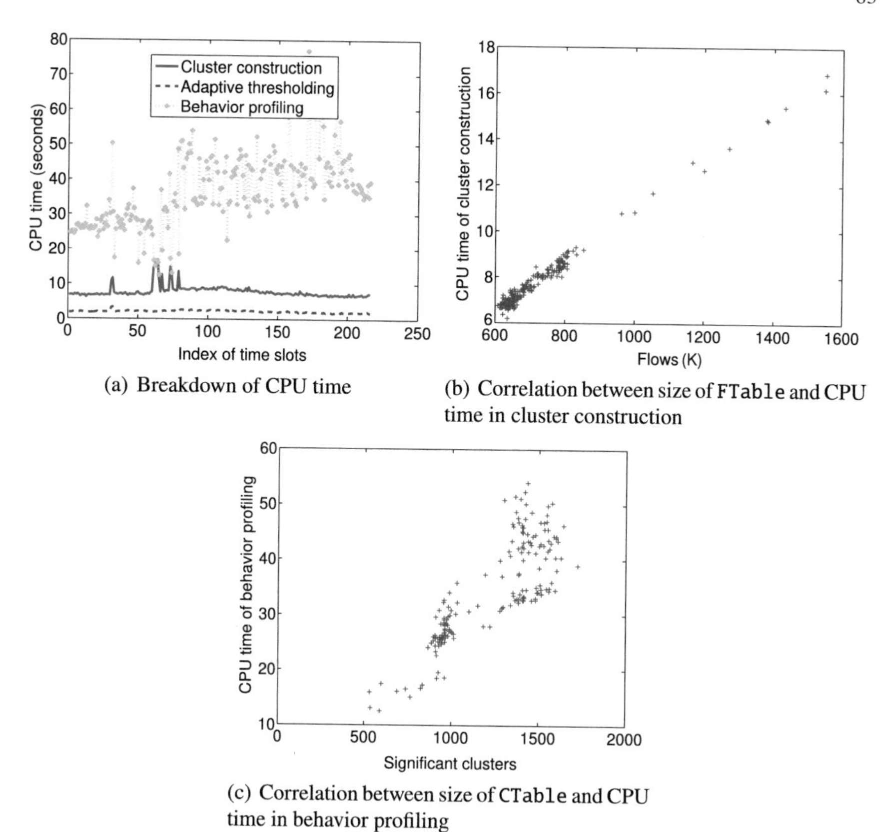

(a) Breakdown of CPU time

(b) Correlation between size of FTable and CPU time in cluster construction

(c) Correlation between size of CTable and CPU time in behavior profiling

Fig. 6.6 Breakdown of CPU time and correlations

Table 6.3 Total CPU and memory cost with various time granularities

Link	Timescale	CPU time (sec)			Memory (MB)		
		min	avg	max	min	avg	max
L_1	1 min	5	12	31	26	36	76
L_1	2 min	10	22	42	31	46	89
L_1	5 min	25	46	65	82	96	183
L_1	10 min	35	82	129	91	151	208
L_1	15 min	45	88	152	145	218	267

our real-time network behavior analysis system satisfies the *scalability* requirement raised in the previous section.

In summary, the average CPU and memory costs of the real-time network behavior analysis system on 5-min flow records collected from an OC48 link with a 10% link utilization are 60 s and 100 MB, respectively. Moreover, the CPU time is largely determined by the number of flow records as well as that of significant clusters, and the memory cost is determined by the number of flow records. During these monitoring periods, these links are not fully utilized, so we cannot extensively measure the performance of the real-time network behavior analysis system for a highly loaded link. Next, we will test the network behavior analysis system during sudden traffic surges such as those caused by denial-of-service attacks, flash crowds, and worm outbreaks that increases the link utilization as well as the number of flow records.

6.3.2 Stress Test

The performance benchmarking of CPU and memory costs demonstrates the operational feasibility of our network behavior analysis system during normal traffic patterns. However, the network behavior analysis system should also be robust during atypical traffic patterns, such as denial-of-service attacks, flash crowds, and worm outbreaks [13–16]. In order to understand the system performance during these incidents, we inject packet traces of three known denial-of-service attacks and simulated worm outbreaks by superposing them with backbone traffic.

We use the packet traces of three DoS attacks with varying intensity and behavior studied in [17]. All of these attacks are targeted on a single destination IP address. The first case is a multiple-source DoS attack, in which hundreds of source IP addresses send 4200 ICMP echo request packets with per second for about 5 min. The second case is a TCP SYN attack lasting 12 min from random IP addresses that send 1100 TCP SYN packets per second. In the last attack, a single source sends over 24K *ip-proto 255* packets per second for 15 min. In addition to DoS attacks, we simulate the SQL slammer worm on January 25, 2003 [14] with an Internet Worm Propagation Simulator used in [13]. In the simulation experiments, we adopt the same set of parameters in [13] to obtain similar worm simulation results and collect worm traffic monitored in a 2^{20} IP space.

For each of these four anomalous traffic patterns, we replay packet traces along with backbone traffic and aggregate synthetic packet traces into 5-tuple flows. For simplicity, we still use 5 min as the size of the time interval and run the network behavior analysis system against the flow records collected in an interval. Table 6.4 shows a summary on flow traces of the first 5-min interval for these four cases. The flow, packet, and byte counts reflect the intensity of attacks or worm propagation, while the link utilization indicates the impact of such anomaly behaviors on Internet links. For all of these cases, the network behavior analysis system is able to successfully generate event reports in less than 5 min.

Table 6.4 Synthetic packet traces with known denial-of-service attacks and worm simulations

Anomaly	Flows	Packets	Bytes	Link utilization	CPU time	Memory	Details
DoS-1	2.08 M	18.9 M	11.8 G	314.5 Mbps	45 s	245.5 MB	DDoS attacks from multiple sources
DoS-2	1.80 M	20.7 M	12.5 G	333.5 Mbps	59 s	266.1 MB	DDoS attacks from random sources
DoS-3	16.5 M	39.8 M	16.1 G	430.1 Mbps	210 s	1.75GB	DoS attacks from single source
Worm	18.9 M	43.0 M	23.6 G	629.2 Mbps	231 s	2.01GB	slammer worm simulations

Table 6.5 Reduction of CPU time and memory cost using the random sampling technique

Case	μ	Size of `FTable`	CPU time	memory
DoS attack	66%	10M	89 s	867 MB
Worm	55%	10M	97 s	912 MB

During the emulation process, the link utilization ranged from 314.5 Mbps to 629.2 Mbps. We run the network behavior analysis system on flow traces after replaying synthetic packets and collect CPU and memory cost of each time interval, which is also shown in Table 6.4. The system works well for low-intense DoS attacks in the first two cases. However, due to intense attacks in the last DoS case (DoS-3) and worm propagations, the CPU time of the system increases to 210 and 231 s, but still under the 5 min interval. However, the memory cost jumps to 1.75 GB and 2.01GB indicating a performance bottleneck. This clearly suggests that we need to provide practical solutions to improve the robustness of the system under stress. In the next section, we will discuss various approaches, including traditional sampling techniques and new profiling-aware filtering techniques towards this problem and evaluate the trade-off between performance benefits and profiling accuracy.

6.4 Sampling and Filtering

However, resource requirements are much higher under anomalous traffic patterns such as sudden traffic surges caused by denial-of-service attacks, when the flow arrival rate can increase by several orders of magnitude. We study this phenomenon by superposing "synthetic" packet traces containing a mix of known denial-of-service (DoS) attacks [17] on real backbone packet traces. To enhance the robustness of our network behavior analysis system under these stress conditions, we propose and develop sampling-based *flow filtering* algorithms and show that these algorithms are able to curb steep increase in CPU and memory costs while maintaining high profiling accuracy.

In this section, we first adopt traditional sampling techniques to address performance bottleneck during sudden traffic surges as caused by severe DoS attacks or worm outbreaks. After evaluating its strength and limitation, we propose a simple yet effective *profiling-aware* filtering algorithm that not only reduces memory cost, but also retains high profiling accuracy.

6.4.1 Random Sampling

Random sampling is a widely used simple sampling technique in which each object, flow in our case, is randomly chosen based on the same probability (also known as sampling ratio μ) [18–20]. Clearly, the number of selected flows is entirely decided by the sampling ratio μ. During the stress test in the last section, the network behavior analysis system requires about 2GB memory when the number of flow records reaches 16.5M and 18.9 during DoS attacks and worm outbreaks. Such high memory requirement is not affordable in real time since the machine installed with the system could have other tasks as well, e.g., packet and flow monitoring. As a result, we attempt to set 1GB as the upper bound of the memory cost. Recall that in the performance benchmarking, we find that memory cost is determined by the number of flow records. Based on their linear relationship shown in Fig. 6.5 we estimate that flow records with a size of 10M will require approximately 1 GB memory. Thus, 10M is the desirable limit for the size of the flow records.

Using the limit of flow records, l, we could configure the sampling ratio during sudden traffic increase as $\mu = \frac{l}{|F|}$. As a result, we set the sampling ratios in the last DoS attacks and worm outbreaks as 60% and 55%, respectively, and randomly choose flows in loading flow tables in the *cluster construction* module. Table 6.5 shows the reduction of CPU time and memory consumptions with the sampled flow tables for both cases.

On the other hand, random sampling has substantial impact on behavior accuracy. First, the set of significant clusters from four feature dimensions are smaller than that without sampling, which is caused by the changes of the underlying cluster size distribution after flow sampling. Table 6.6 shows the number of significant clusters

Table 6.6 Reduction of significant clusters and behavior accuracy

Dim.	Sig. clusters without sampling	Sig. clusters with sampling	Clusters with same behavior classes
srcPrt	23	4	3
dstPrt	6	5	4
srcIP	47	31	29
dstIP	233	140	125
Total	309	180	161

extracted along each dimension without and with sampling for the DoS case. In total, among 309 significant clusters without sampling, 180 (58%) of the *most significant* clusters are still extracted with random sampling. Secondly, the behavior of a number of extracted clusters is altered, since flow sampling changes the feature distribution of free dimensions as well as the behavior classes for these clusters. As shown in the last column of Table 6.6, 161 out of 180 significant clusters with random sampling are classified with the same behavior as those without sampling. In other words, the behavior of 19 (10.5%) extracted significant clusters has changed as a result of random sampling. Figure 6.7 shows the feature distributions of free dimensions for 140 `dstIP` clusters with and without random sampling. The deviations from the diagonal line indicate the changes of feature distribution and the behavior due to flow sampling. We also perform random sampling on the synthetic flow traces in the case of worm outbreak, and the results of sampling impact on cluster extractions and behavior accuracy are very similar.

In summary, random sampling could reduce the CPU time and memory cost during sudden traffic surges caused by DoS attacks or worm outbreaks. However, random sampling reduces the number of interesting events, and also alters the behavior classes of some significant clusters. Such impact could have become worse if "lower" sampling rates are selected. Thus, it becomes necessary to develop a profiling-aware algorithm that not only reduces the size of flow tables, but also retains the (approximately) same set significant clusters and their behavior.

6.4.2 *Profiling-Aware Filtering*

A key lesson from random sampling is that the clusters associated with denial-of-service attacks are usually very large in flow count, and hence consume a large amount of memory and CPU time [21]. In addition, profiling such behavior does not require a large number of flows, since the feature distributions very likely remain the same even with a small percentage of traffic flows. Based on this insight, we develop a profiling-aware filtering solution that limits the size of very large clusters,

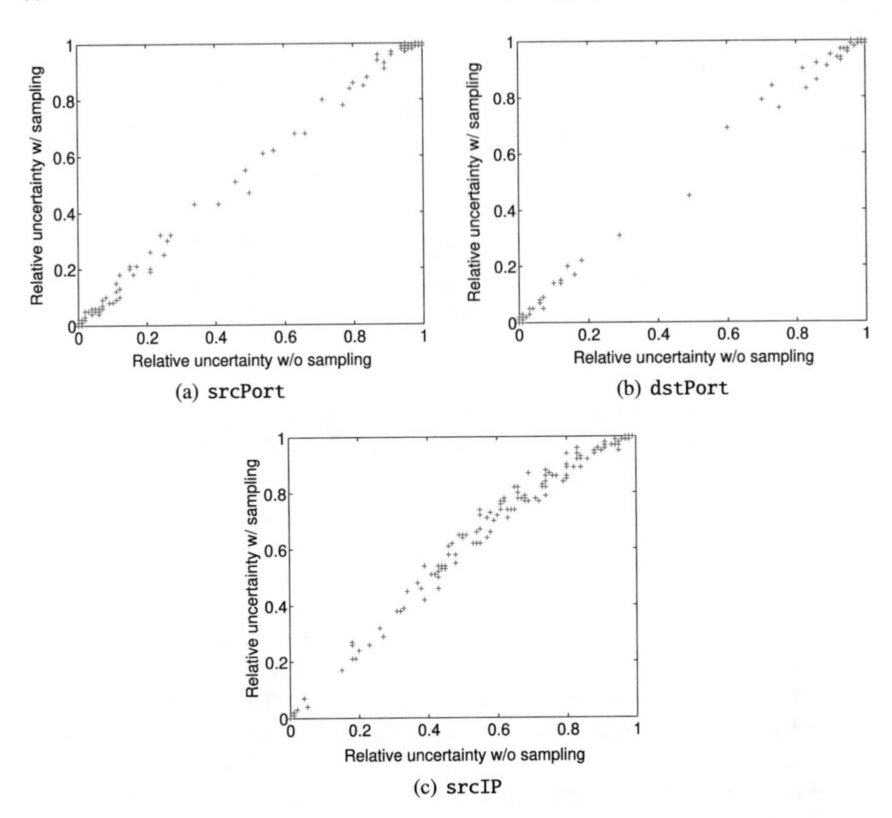

Fig. 6.7 Feature distribution of free dimensions for 140 `dstIP` clusters with and without random sampling

and adaptively samples on the rest of clusters when the system is faced with sudden explosive growth in the number of flows.

Algorithm 1 describes the details of the profiling-aware sampling algorithm. First, we choose two *watermarks* (L and H) for the network behavior analysis system. L represents the moving average of flow tables over time, and H represents the maximum size of flow tables that system will accept. In our experiments, we set $H = 10M$, which is estimated to require 1GB memory cost. In addition, we set the maximum and minimum sampling ratios, i.e., μ_{max} and μ_{min}. The actual sampling μ will be adaptively decided based on the status of flow table size. Specifically, the sampling ratio becomes thinner as the size of flow table increases. For simplicity, let `ftable` denote the size of flow table. If `ftable` is below L, the network behavior analysis system accepts every flow. In contrary, if `ftable` is equal to H, the system will stop reading flows and exit with a warning signal.

If `ftable` is equal to L or certain marks, i.e., $L + i * D$, where D is the incremental factor and $i = 1, 2..., (H - L)/D - 1$, the system computes the relative uncertainty of each dimension and evaluates whether there is one or a few dominant feature

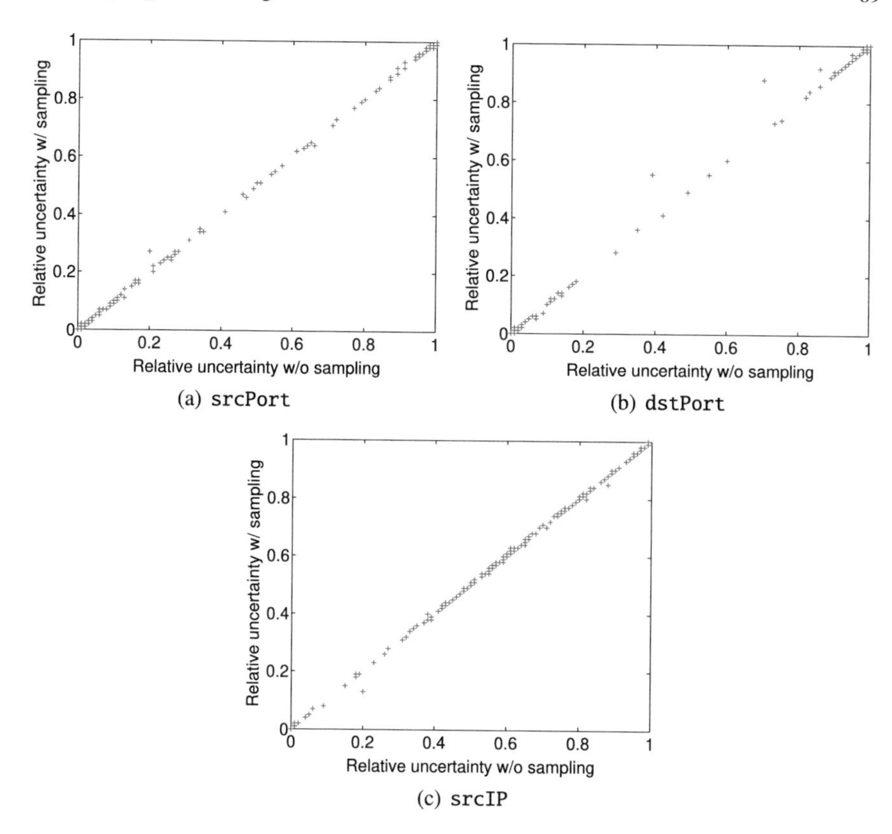

Fig. 6.8 Feature distribution of free dimensions for 210 `dstIP` clusters with and without profiling-aware sampling

values along each dimension. In our experiments, we set $D = 1M$ as the incremental factor. The existence of such values suggests that certain types of flows dominate current flow tables and indicates anomalous traffic patterns. Thus, the system searches these values and marks them as significant clusters for flow filtering. Subsequently, any flow, which contains a feature value marked with *significant*, will be filtered, since such flow will not affect the behavior of the associated clusters. On the other hand, additional flows for other small clusters have substantial contributions to their behavior. Thus, we should give preference to flows that belong to such small clusters. On the other hand, the system could not accept all of these flows with preference after $ftable$ exceeds L watermark. As a result, each of these flows is added with the adaptive sampling ratio

$$\mu = \mu_{max} - i * \frac{\mu_{max} - \mu_{min}}{(H - L)/D - 1}. \tag{6.9}$$

Algorithm 3 A Profiling-aware filtering algorithm

```
1: Parameters: L, H, D, μ_max, μ_min, β*
2: Initilization: I = (H − L)/D
   δ_μ = (μ_max − μ_min)/I
   μ = μ_max
   i = 0
   ftable = 0
3: srcPrt = 0; dstPrt = 1
4: srcIP = 2; dstIP = 3
5: while next flow do
6:   if (ftable < L) then
7:     Insert flow into FTable (Flow Table)
8:     ftable ++
9:     continue;
10:  else
11:    if (ftable > H) then
12:      Stop reading flows
13:      exit
14:    end if
15:  end if
16:  if (ftable == L + i * D) then
17:    ru_0 = Relative_Uncertainty(FTable, srcPrt)
18:    ru_1 = Relative_Uncertainty(FTable, dstPrt)
19:    ru_2 = Relative_Uncertainty(FTable, srcIP)
20:    ru_3 = Relative_Uncertainty(FTable, dstIP)
21:    for (dim = 0; dim ≤ 3; dim ++) do
22:      ru = ru_dim
23:      while (ru ≤ β*) do
24:        remove feature value with highest probability
25:        mark feature value as significant
26:        ru = Relative_Uncertainty(FTable, dim)
27:      end while
28:    end for
29:    μ = μ_max − i * δ_μ
30:    i ++
31:  end if
32:  if ((ftable ≥ L) && (ftable ≤ H)) then
33:    if (any feature value in flow is marked significant) then
34:      Filter flow
35:    else
36:      Insert flow into FTable with probability μ
37:      if (flow is selected then
38:        ftable ++
39:      end if
40:    end if
41:  end if
42: end while
```

We run the network behavior analysis system on the flow tables in the cases of DoS attack and worm outbreaks (cf. Table 6.6) with the profile-aware filtering algorithm. Like random sampling, our *profiling-aware* sampling solution also reduces CPU time and memory cost by limiting the size of flow table. On the other hand, the profiling-aware sampling has two advantages over the random sampling. First, the set of clusters extracted using this algorithm is very close to the set without sampling. For example, in the case of DoS attack, the system obtains 41 `srcIP` clusters, 210 `dstIP` clusters, 21 `srcPrt` clusters, and 6 `dstPrt` clusters, respectively. Compared with 58% of significant clusters extracted in random sampling, our profiling-aware algorithm could extract over 90% of 309 original clusters that are selected without any sampling. Second, the behavior accuracy of significant clusters is also improved. Specifically, among 41 `srcIP`'s, 210 `dstIP`'s, 21 `srcPrt`'s, and 6 `dstPrt`'s significant clusters, only 3 `dstIP`'s and 1 `srcPrt` clusters change to "akin" classes from their original behavior classes. These findings suggest that the *profiling-aware* profiling algorithm approximately retains the feature distributions of significant clusters and behaviors.

Figure 6.8 shows the feature distribution of free dimensions of 210 `dstIP` clusters, extracted both without sampling and with profiling-aware filtering algorithm. In general, the feature distributions of all free dimensions for almost all clusters after filtering are approximately the same as those without sampling. The outliers deviant from the diagonal lines correspond to feature distributions of three clusters whose behavior has changed. Upon close examinations, we find that flows in these clusters contain a mixture of Web and ICMP traffic. The latter are the dominant flows in DoS attacks, so they are filtered after the size of flow table reaches L in the profiling-aware filtering algorithm. The filtered ICMP flows in these clusters explain the changes of the feature distributions as well as the behavior classes.

In the *worm* case, the profiling-aware filtering algorithm also successfully reduces CPU and memory cost of the network behavior analysis system, while maintaining high profiling accuracy in terms of the number of extracted significant clusters and the feature distributions of these clusters. Thus, the profiling-aware filtering algorithm can achieve a significant reduction of CPU time and memory cost during anomalous traffic patterns while obtaining accurate behavior profiles of most networked systems and Internet applications.

6.5 Summary

This chapter explores the feasibility of designing, implementing, and utilizing a real-time network behavior analysis system for high-speed Internet links. We first discuss the design requirements and challenges of such a system and present an overall architecture that integrates the network behavior analysis system with always-on monitoring systems and an event analysis engine. Subsequently, we demonstrate the operational feasibility of building this system through extensive performance benchmarking of CPU and memory costs using a variety of packet traces collected from

OC48 backbone links. To improve the robustness of this system during anomalous traffic patterns such as denial-of-service attacks or worm outbreaks, we propose a simple yet effective filtering algorithm to reduce resource consumptions while retaining high behavioral profiling accuracy.

References

1. G. Iannaccone, CoMo: An Open Infrastructure for Network Monitoring - Research Agenda. Tech. rep., Intel Research Technical Report (2005)
2. G. Iannaccone, C. Diot, I. Graham, N. McKeown, Monitoring very high speed links, in *Proceedings of ACM SIGCOMM Internet Measurement Workshop* (2001)
3. K. Keys, D. Moore, C. Estan, A robust system for accurate real-time summaries of internet traffic, in *Proceedings of ACM SIGMETRICS* (2005)
4. K. Xu, F. Wang, S. Bhattacharyya, Z.-L. Zhang, A real-time network traffic profiling system, in *Proceedings of International Conference on Dependable Systems and Networks* (2007)
5. K. Xu, F. Wang, S. Bhattacharyya, Z.-L. Zhang, Real-time behavior profiling for network monitoring. Int. J. Internet Protoc. Technol. **5**(1/2), 65–80 (2010)
6. A. Lakhina, M. Crovella, C. Diot, Mining anomalies using traffic feature distributions, in *Proceedings of ACM SIGCOMM* (2005)
7. K. Xu, Z.-L. Zhang, S. Bhattacharyya, Profiling internet backbone traffic: behavior models and applications, in *Proceedings of ACM SIGCOMM* (2005)
8. H. Kim, B. Karp, Autograph: toward automated, distributed worm signature detection, in *Proceedings of USENIX Security Symposium* (2004)
9. J. Newsome, B. Karp, D. Song, Polygraph: automatic signature generation for polymorphic worms, in *Proceedings of IEEE Security and Privacy Symposium* (2005)
10. S. Singh, C. Estan, G. Varghese, S. Savage, Automated worm fingerprinting, in *Proceedings of Symposium on Operating Systems Design and Implementation* (2004)
11. A. Chakrabarti, K. Ba, S. Muthukrishnan, Estimating entropy and entropy norm on data streams, in *Proceedings of Symposium on Theoretical Aspects of Computer Science (STACS)* (2006)
12. A. Lall, V. Sekar, M. Ogihara, J. Xu, H. Zhang, Data streaming algorithms for estimating entropy of network traffic, in *Proceedings of ACM SIGMETRICS* (2006)
13. C. Zou, L. Gao, W. Gong, D. Towsley, Monitoring and early warning for internet worms, in *Proceedings of ACM Conference on Computer and Communications Security* (2003)
14. D. Moore, V. Paxson, S. Savage, C. Shannon, S. Staniford, N. Weaver, Inside the Slammer Worm. IEEE Security and Privacy (2003)
15. J. Jung, B. Krishnamurthy, M. Rabinovich, Flash crowds and denial of service attacks: characterization and implications for CDNs and web sites, in *Proceedings of International World Wide Web Conference* (2002)
16. S. Kandula, D. Katabi, M. Jacob, A. Berger, Botz-4-Sale: surviving organized DDoS attacks that mimic flash crowds, in *Proceedings of Symposium on Networked Systems Design and Implementation* (2005)
17. A. Hussain, J. Heidemann, C. Papadopoulos, A framework for classifying denial of service attacks, in *Proceedings of ACM SIGCOMM* (2003)
18. B.-Y. Choi, J. Park, Z.-L. Zhang, Adaptive random sampling for load change detection, in *Proceedings of ACM SIGMETRICS* (2002)
19. K. Claffy, H.-W. Braun, G. Polyzos, Application of sampling methodologies to wide-area network traffic characterization, in *Proceedings of ACM SIGCOMM* (1993)
20. N. Duffield, C. Lund, M. Thorup, Charging from sampled network usage, in *Proceedings of ACM SIGCOMM Internet Measurement Workshop* (2001)
21. K. Argyraki, D. Cheriton, Active internet traffic filtering: real-time response to denial-of-service attacks, in *Proceedings of USENIX Annual Technical Conference* (2005)

Chapter 7
Applications

Abstract Network behavior analysis provides critical insights and visibility on what is happening to millions of networked systems and thousands of Internet applications in a variety of network environments. A number of studies have demonstrated the benefits and feasibility of these behavioral insights and visibility in a wide spectrum of applications such as traffic profiling, cybersecurity, and network forensics. This chapter first presents the applications of network behavior analysis in profiling Internet traffic and discovering server and service behavior profiles, heavy-hitter host behavior profiles, scan and exploit profiles, and deviant or rare behavior profiles. Subsequently, this chapter discusses how network behavior analysis enhances cybersecurity by discovering and stopping scanning and exploit traffic from the Internet. Finally, this chapter sheds light on the applications of cluster-aware network behavior analysis by exploring the benefits of end-host behavior clusters and application behavior clusters, particularly in characterizing traffic patterns of network prefixes and detecting emerging applications and threats, which share strong similarities with existing and known applications and threats.

7.1 Profiling Internet Traffic

We apply behavior modeling and structural modeling to obtain general profiles of the Internet backbone traffic based on the datasets listed in Table 3.1. We find that a large majority of the (significant) clusters fall into three "canonical" profiles: typical *server/service behavior* (mostly providing well-known services), typical *"heavy-hitter" host behavior* (predominantly associated with well-known services), and typical *scan/exploit behavior* (frequently manifested by hosts infected with known worms). The canonical behavior profiles are characterized along the following four key aspects: (i) BCs they belong to and their properties, (ii) temporal characteristics (frequency and stability) of individual clusters, (iii) dominant states, and (iv) additional attributes such as average flow size in terms of packet and byte counts and their variabilities.

Clusters with behaviors that differ in one or more aspects of the three canonical profiles automatically present themselves as more interesting, thus warrant closer

K. Xu, *Network Behavior Analysis*,
https://doi.org/10.1007/978-981-16-8325-1_7

Table 7.1 Three canonical behavior profiles

Profile	Dimension	BCs	Examples
Servers	srcIP	$BC_{6,7,8}$	web, DNS, email
or	dstIP	$BC_{18,19,20}$	
Services	srcPrt	BC_{23}	aggregate service
	dstPrt	BC_{25}	traffic
Heavy	srcIP	$BC_{18,19}$	NAT boxes
Hitter Hosts	dstIP	$BC_{6,7}$	web proxies, crawlers
Scans	srcIP	$BC_{2,20}$	scanners, exploits
or	dstIP	$BC_{2,8}$	scan targets
Exploits	dstPrt	$BC_{2,5,20,23}$	aggregate exploit traffic

examination. Generally speaking, there are two types of *interesting* or *anomalous* behaviors we find using our behavior profiling methodology: (i) novel or unknown behaviors that match the typical server/service profile, heavy-hitter host profile, or scan/exploit profile, but exhibit *unusual* feature values, as revealed by analyses of their dominant states; and (ii) deviant or abnormal behaviors that deviate significantly from the canonical profiles in terms of BCs (e.g., clusters belonging to rare BCs), temporal instability (e.g., unstable clusters that jump between different BCs), or additional features.

7.1.1 Server/Service Behavior Profiles

As shown in Table 7.1, a typical server providing a well-known service shows up in either the popular, large and non-volatile `srcIP` BC_6 [0,2,0], BC_7 [0,2,1] and BC_8 [0,2,2], or `dstIP` BC_{18} [2,0,0], BC_{19} [2,0,1] and BC_{20} [2,0,2] (note the symmetry between the `srcIP` and `dstIP` BCs, with the first two labels (`srcPrt` and `dstPrt`) swapped). These BCs represent the behavior patterns of a server communicating with a few, many or a large number of hosts. In terms of their temporal characteristics, the individual clusters associated with servers/well-known services tend to have a relatively high frequency, and almost all of them are stable, re-appearing in the same or akin BCs. The average flow size (in both packet and byte counts) of the clusters shows high variability, namely each cluster typically consists of flows of different sizes.

Looking from the `srcPrt` and `dstPrt` perspectives, the clusters associated with the well-known service ports almost always belong to the same BC's, e.g., either `srcPrt` BC_{23} [2,1,2] or `dstPrt` BC_{25} [2,2,1], representing the aggregate behavior of a (relatively smaller) number of servers communicating with a much larger number of clients on a specific well-know service port. For example, Fig. 7.1a

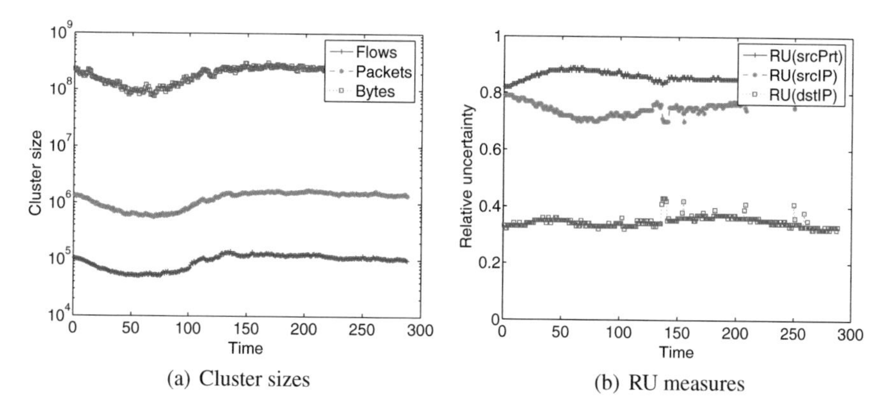

Fig. 7.1 Cluster sizes (in flow, packet, and byte counts) and RU measures of the `dstPrt` 80 cluster (aggregate web traffic) on L_1 over time

plots the *cluster* sizes (in flow, packet, and byte counts) of the `dstPrt` TCP 80 cluster (representing aggregate behavior of all web servers) over the 24-h period, whereas in Fig. 7.1b, we plot the corresponding RU_{srcPrt}, RU_{srcIP}, and RU_{dstIP} of its three free dimensions over time. We see that the `dstPrt` TCP port 80 cluster is highly persistent, observed in every time slot over the 24-h period, with the number of `srcIP`'s (web servers) fairly stable over time. The cluster size over time shows a diurnal pattern, but otherwise does not fluctuate dramatically. In addition, the packet and byte counts of the cluster are considerably larger than the total number of flows, indicating that on the average each flow contains at least several packets and a few hundred bytes.

An overwhelming majority of the `srcIP` clusters in $BC_{6,7,8}$ are corresponding to Web, DNS or Email servers. They share very similar behavior characteristics, belonging to the same BC's, stable with relatively high frequency, and containing flows with diverse packet/byte counts. Among the remaining clusters, most are associated with http-alternative services (e.g., 8080), https(443), real audio/video servers (7070), IRC servers (6667), and peer-to-peer (P2P) servers (4662). Most interestingly, we find three `srcIP` clusters with service ports 56192, 56193, and 60638. They share similar characteristics with web servers, having a frequency of 12, 9, and 22, respectively, and with diverse flow sizes both in packet and byte counts. These observations suggest that they are likely servers running on unusual high ports. Hence, these cases represent examples of "novel" service behaviors that our profiling methodology is able to uncover.

7.1.2 Heavy-Hitter Host Behavior Profiles

The second canonical behavior profile is what we call the *heavy-hitter* host profile [1, 2], which represents hosts (typically clients) that send a large number of flows to a single or a few other hosts (typically servers) in a short period of time (e.g., a 5-min period). They belong to either the popular and non-volatile `srcIP` BC_{18} [2,0,0] or BC_{19} [2,0,1], or the `dstIP` BC_6 [0,2,0] and BC_7 [0,2,1]. The frequency of individual clusters is varied, with a majority of them having medium frequency, and almost all of them are stable. These heavy-hitter clusters are typically associated with well-known service ports (as revealed by the dominant state analysis), and contain flows with highly diverse packet and byte counts. Many of the heavy-hitter hosts are corresponding to NAT boxes (many clients behind a NAT box making requests to a few popular web sites, making the NAT box a heavy-hitter), web proxies, cache servers, or web crawlers.

For example, we find that 392 and 429 unique `srcIP` clusters from datasets L_1 and L_2 belong to BC_{18} and BC_{19}. Nearly 80% of these heavy-hitters occur in at least five time slots, exhibiting consistent behavior over time. The most frequent ports used by these hosts are TCP port 80 (70%), UDP port 53 (15%), TCP port 443 (10%), and TCP port 1080(3%). However, there are heavy-hitters associated with other rarer ports. In one case, we found one `srcIP` cluster from a large corporation talking to one `dstIP` on TCP port 7070 (RealAudio) generating flows of varied packet and byte counts. It also has a frequency of 11. Deeper inspection reveals this is a legitimate proxy, talking to an audio server. In another case, we found one `srcIP` cluster talking to many `dstIP` hosts on TCP port 6346 (Gnutella P2P file sharing port), with flows of diverse packet and byte counts. This host is thus likely a heavy file downloader. These results suggest that the profiles for heavy-hitter hosts could be used to identify these unusual heavy-hitters.

7.1.3 Scan/Exploit Profiles

Behaviors of hosts performing scans or attempting to spread worms or other exploits constitute the third canonical profile. Two telling signs of typical scan/exploit behavior [3] are (i) the clusters tend to be highly volatile, appearing and disappearing quickly, and (ii) most flows in the clusters contain one or two packets with fixed size, albeit occasionally they may contain three or more packets (e.g., when performing OS fingerprinting or other reconnaissance activities). For example, we observe that most of the flows using TCP protocol in these clusters are failed TCP connections on well-known exploit ports. In addition, most flows using UDP protocol or ICMP protocol have a fixed packet size that matches widely known signature of exploit activities, e.g., UDP packets with 376 bytes to destination port 1434 (Slammer Worm), ICMP packets with 92 bytes (ICMP ping probes). These findings provide additional evi-

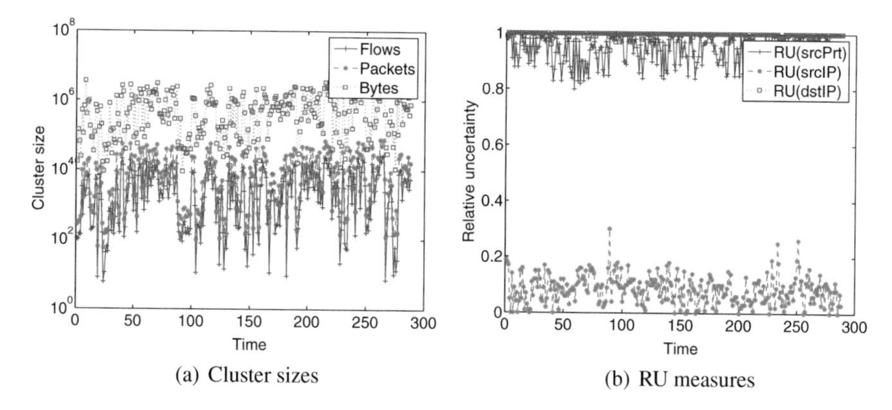

(a) Cluster sizes (b) RU measures

Fig. 7.2 Cluster sizes and RU measures of the dstPrt(3127) cluster (aggregate myDoom traffic) on L_2 over time. Note that in **a** the lines of flow count and packet count are indistinguishable, since most flows in the clusters contain a singleton packet

dence to confirm that such clusters are likely associated with scanning or exploit activities.

A disproportionately large majority of extracted clusters fall into this category, many of which are among the top in terms of flow counts (but in general, not in byte counts, cf. Fig. 4.2). Such prevalent behavior signifies the severity of worm/exploit spread and the magnitude of infected hosts (cf. [4, 5]). On the plus side, however, these hosts manifest distinct behavior that is clearly separable from the server/service or heavy-hitter host profiles: the srcIP clusters (a large majority) belong to BC_2 [0,0,2] and BC_{20} [2,0,2], corresponding to hosts performing scan or spreading exploits to random dstIP hosts on a fixed dstPrt using either fixed or random srcPrt's; the dstIP clusters (a smaller number) belong to BC_2 [0,0,2] and BC_8 [0,2,2], reflecting hosts (victims of a large number of scanners or attacks) responding to probes on a targeted srcPrt. Using specific dstPrt's that are targets of known exploits, e.g., 1434 (used by SQL Slammer), the aggregate traffic behavior of exploits is also evidently different from that of normal service traffic behavior (e.g., web): the dstPrt clusters typically belong to BC_{23} [2,1,2], but sometimes to BC_2 [0,0,2], BC_5 [0,1,2], or BC_{20} [2,0,2], representing a relatively smaller number of srcIP hosts probing a larger number of dstIP hosts on the target dstPrt using either fixed or random srcPrt's. This is in stark contrast with normal service traffic aggregate such as web (i.e., dstPrt 80 cluster), where a much larger number of clients (srcIP's) talk to a relatively smaller number of servers (dstIP's) using randomly generated srcPrt's, thus belonging to dstPrt BC_{25} [2,2,1].

To contrast the aggregate exploit behavior with the aggregate service profile shown in Fig. 7.1, we examine cluster sizes of the dstPrt 3127 cluster (aggregate myDoom traffic), and the corresponding RU measures of the three free dimensions over time. Unlike Web traffic, we see that the aggregate traffic fluctuates a lot over time and the flow count and packet count are very similar, indicating flows with a few packets. It is important to note that the dstPrt(3127) cluster is considered "significant" by Algorithm 1 in 132 time slots out of the total 288. For ease of comparison, we also extracted the flows with dstPrt(3127) in the other time slots when it is not significant. On the other hand, such traffic exhibits low RU_{srcIP} (mostly below 0.1) and high RU_{dstIP} (≈ 1.0) over time, indicating that a few sources (same or different) are randomly probing a large number of targets on and off over time (Fig. 7.2).

In addition to those dstPrt's that are known to have exploits, we also find several (srcIP) clusters that manifest typical scan/exploit behavior, but are associated with dstPrt's that *we do not know* to have known exploits. For example, we find that in one time slot a srcIP cluster is probing a large number of destinations on UDP port 12827, with a single UDP packet. This host could simply engage in some harmless scanning on UDP port 12827, but it could also be a new form of RATs (remote access trojans) or even a precursor of something more malicious. Further inspection is clearly needed. Nonetheless, it illustrates that our profiling technique is capable of automatically picking out clusters that fit the scan/exploit behavior profile but with unknown feature values. This will enable network operators/security analysts to examine novel, hitherto unknown, or "zero-day" exploits.

We evaluate exploit behavior profiles using a set of heuristics that do not require packet payload. These heuristics are derived from the domain knowledge in network protocols such as TCP, UDP, and ICMP, and some of them have been used in other studies [6, 7] as well.

We first study TCP flows from srcIP's with exploit profiles. To establish a successful TCP connection, a srcIP and a dstIP must go through the three-way handshake process by exchanging SYN, SYN/ACK, and ACK packets. Thus, the initiator or responder send at least three or two packets when considering the data packets. In contrast, in failed TCP connections, the initiator sends less than three packets. Given this insight, we examine all 12.9M TCP flows sent by srcIP's with exploit profiles in L_1. The results show that 97% of them have exactly one packet and 1.8% of them have two packets. Thus, these flows are likely failed TCP connections to exploit ports. The established TCP connections correspond to legitimate traffic flows from these srcIPs or successful exploit activities.

For flows with UDP or ICMP packets, we examine the packet size of each flow since many well-known exploits (e.g., SQL Slammer Worm) or probing activities (e.g., ICMP ping) have a fixed packet size. We separate all UDP flows from srcIPs with exploit profile based on dstPrt. To illustrate our finding, we study two popular exploit ports, UDP 1434 and UDP 137 using L_1. The results show that 100% of flows from srcIP's with exploit profiles on UDP 1434 have a fixed packet size of 404 bytes, while 99.8% of flows from them on UDP 137 have a fixed packet size of 78 bytes. The byte counts are part of widely known exploit signatures of SQL Slammer worm and UDP 137 scan activity.

For flows with ICMP packets, we examine the flows with 92 bytes that are generated by ICMP ping. Using the packet traces of L_1, we find that among 7.2M ICMP flows sent by 897 srcIP's with exploit profiles, 97.2% of flows are a single packet of size 92 bytes. Thus, these srcIP's were likely probing the destinations.

These evaluations demonstrate that our profiling methodology is effective to capture typical and anomalous communication patterns. In addition, we also analyze how many hosts with exploit behaviors are not profiled. Since our focus is on significant communication patterns of end hosts or service, it is not surprising that many hosts and service are not profiled in every 5-min time period. For example, by examining the packet traces of L_1 over 24 h, we find that our methodology captures less than 25% of all `srcIP`'s that send packets on TCP port 135 in each period. However, these significant `srcIP`'s extracted account for about 80% of all flows on TCP port 135 over time [3]. In other words, the methodology builds exploit behavior profiles for a few `srcIP`'s, but with a substantial amount of exploit traffic.

7.1.4 Deviant or Rare Behaviors

We have demonstrated how we are able to identify novel or anomalous behaviors that fit the canonical profiles but contain unknown feature values (as revealed by the dominant state analysis). We now illustrate how rare behaviors or deviant behaviors are also indicators of anomalies, and thus worthy of deeper inspection. In the following, we present a number of case studies, each of which is selected to highlight a certain type of anomalous behavior. Our goal here is not to exhaustively enumerate all possible deviant behavioral patterns, but to demonstrate that building a comprehensive traffic profile can lead to the identification of such patterns.

Clusters in rare behavior classes. The clusters in the rare behavior classes by definition represent atypical behavioral patterns. For example, we find three `dstPrt` clusters (TCP ports 6667, 113 and 8083) suddenly appear in the rare `dstPrt` BC_{15} [1,2,0] in several different time slots, and quickly vanish within one or two time slots. Close examination reveals that more than 94% of the flows in the clusters are destined to a single `dstIP` from random `srcIP`'s. The flows to the dstIP have the same packet and byte counts. This evidence suggests that these `dstIP`'s are likely experiencing a DDoS attack.

Behavioral changes for clusters. Clusters that exhibit unstable behaviors such as suddenly jumping between BCs (especially when a frequent cluster jumps from its usual BC to a different BC) often signify anomalies. In one case, we observe that one `srcIP` cluster (a popular web server) on L_1 makes a sudden transition from BC_8 to BC_6, and then moves back to BC_8. Before the transition, the server is talking to a large number of clients with diverse flow sizes. After the behavior transition to BC_6, a single `dstIP` accounts for more than 87% of the flows, and these flows all have the same packet and byte counts. The behavior of the particular client is suspicious. In another case, we find that the `dstPrt` 25 cluster (aggregate email traffic) shifts from its usual BC_{25} [2,2,1] to BC_{23} [2, 1, 2] for three consecutive time slots, a sign of significant changes in the underlying distribution of `srcIP`'s and `dstIP`'s. Upon deeper investigation, we find that this is the result of a single `srcIP` scanning 45,000 random `dstIP`'s on SMTP port using 48-byte packets. This example illustrates how fundamental shifts in communication patterns can point a network security analyst to genuinely suspicious activities.

Unusual profiles for popular service ports. Clusters associated with common service ports that exhibit behaviors that do not fit their canonical profiles are of particular concern, since these ports are typically not blocked by firewalls. For example, we have found quite a few `srcIP` clusters in BC_2 and BC_{20} that perform scans on `dstPrt` 25, 53, 80, etc. Similar to the clusters with known exploit ports, these `srcIP` clusters have small packet and byte counts with very low variability. Note that these common service ports are generally used by a very large number of clients, thereby making it impossible to examine the behavior of each client individually. Our profiling technique, however, can automatically separate out a handful of potentially suspicious clients that use these ports for malicious activities. Lastly, we want to comment that many of these "anomalous" behaviors do not fall into the clusters with, say 1% or 2% of the total number of either flow, packet or byte counts in the traffic. Hence, focusing only on top N clusters in one or multiple dimensions may miss some interesting, anomalous behaviors.

To summarize, we have demonstrated the applicability of the behavioral profiling methodology to critical problem of detecting anomalies or the spread of unknown security exploits, profiling unwanted traffic and tracking the growth of new applications. By applying the behavioral profiling methodology on traffic data collected from a variety of links at the core of the Internet through *off-line* analysis, we find that a large fraction of clusters fall into three typical behavior profiles: server/service behavior profile, heavy hitter host behavior, and scan/exploit behavior profile. These behavior profiles are built based on various aspects, including behavior classes, dominant states, and additional attributes such as average packets and bytes per flow. The behavioral methodology is able to find various interesting and anomalous events. First, it automatically detects novel or unknown exploit behaviors that match typical exploit profiles, but exhibit unusual dominant states (e.g., `dstPrt`'s). Second, any atypical behavior is worth close examination, since they represent as outliers or anomaly among behavior profiles. Third, the methodology could point out deviant behaviors of end hosts or applications that deviate from previous patterns.

7.2 Reducing Unwanted Traffic on the Internet

Recently, we have seen a tremendous increase in unwanted or exploit traffic—malicious or unproductive traffic that attempts to compromise vulnerable hosts, propagate malware, spread spam, or deny valuable services [4, 5, 8–11]. A significant portion of this traffic is due to self-propagating worms, viruses, or other malware; this leads to a vicious cycle as new hosts are infected, generating more unwanted traffic, and infecting other vulnerable hosts. In addition to self-propagating malware, new variants of old malware or new exploits emerge faster than ever, producing yet more unwanted traffic. Strictly speaking, *exploit* traffic means network traffic that is generated with the explicit intention to exploit certain vulnerabilities in target systems—a large subset of *unwanted* traffic. However, these two terms are often used interchangeably in the literature.

Current measures in stopping or reducing unwanted or exploit traffic rely on various firewalls or similar devices deployed on the *end hosts* or at *stub networks* (i.e., networks such as enterprise or campus networks that do not provide *transit* services) to block such traffic. In this chapter, we explore the feasibility and effectiveness of stopping or reducing unwanted traffic from the perspective of an IP backbone network based on our behavior profiling framework.

Given the exploit traffic identified in our traffic profiling methodology, we consider blocking strategies an ISP may pursue to reduce unwanted traffic, by installing access control lists (ACLs) on routers at entry points of an ISP. Although most of exploit traffic is associated with a relatively small set of (destination) ports, simply blocking these ports from any source is, in general, infeasible for a backbone ISP. This is because many ports that are vulnerable to attacks such as port 1434 (Microsoft SQL server) [12] or port 139 (Common Internet File System for Windows) are also used by legitimate applications run by an ISP's customers. An alternate approach is to block the specific offending sources (and the exploit destination ports) of exploit traffic. However, these sources can number in tens or hundreds of thousands for a large backbone network; hence, there is a significant scalability problem (primarily due to overheads incurred in backbone routers for filtering traffic using ACLs) in attempting to block each and every one of these sources. Hence, this approach is likely to be most cost-effective when used to block the top offending sources that send a majority of self-propagating exploit traffic, in particular, in the early stage of a malware outbreak, to hinder their spread.

7.2.1 Unwanted Exploit Traffic on the Internet

As discussed in previous chapters, normal server and client behavior profiles have distinct behavior from scan or exploit profiles. Figure 7.3a illustrates the relative uncertainty vectors of normal server and client behavior, while Fig. 7.3b illustrates those of exploit profiles. The points on the left side of Fig. 7.3a belong to the server profile, where they share a strong similarity in RU_{srcPrt} (low uncertainty) and RU_{dstPrt} (high uncertainty): a server typically talks to many clients using the same service `srcPrt` and randomly selected `dstPrt`'s. The cluster on the right side of Fig. 7.3a belong to the heavy-hitter profile, where they share a strong similarity in RU_{srcPrt} (high uncertainty), RU_{dstPrt} (low uncertainty), and have *low-to-medium* uncertainty in RU_{dstIP}: a heavy-hitter client host tends to talk to a limited number of servers using randomly selected `srcPrt`'s but the same `dstPrt`. Closer inspection reveals that the `srcPrt`'s in the server profile almost exclusively are the well-known service ports (e.g., TCP port 80); whereas the majority of the `dstPrt`'s in the heavy-hitter profile are the well-known service ports, but they also include some popular peer-to-peer ports (e.g., TCP port 6346).

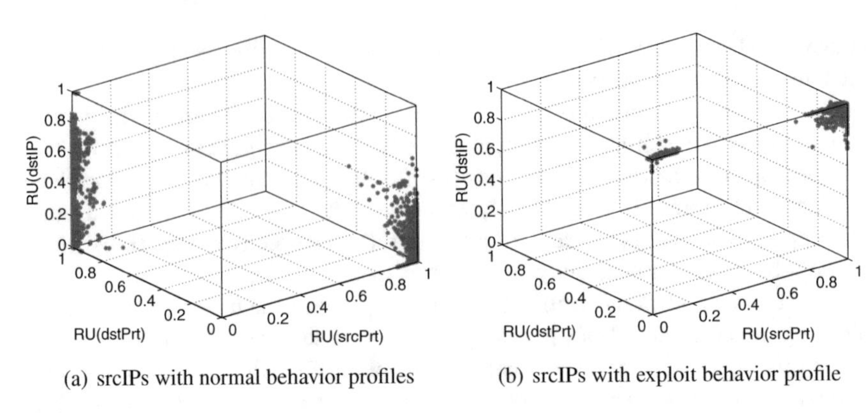

(a) srcIPs with normal behavior profiles (b) srcIPs with exploit behavior profile

Fig. 7.3 The RU vector distribution of the canonical behavior profiles for significant `srcIP`'s in L_1 during a 24-h period

In contrast, the points in the exploit traffic profile (Fig. 7.3b) all have high uncertainty in RU_{dstIP} and low uncertainty in $RU_{dstPort}$, and fall into two categories in terms of $RU_{srcPort}$. Closer inspection reveals that the `dstPorts` include various known exploit ports (e.g., TCP ports 135, 137, 138, 445, UDP ports 1026-28) as well as a few high ports with unknown vulnerabilities. They also include some well-known service ports (e.g., TCP 80) as well as ICMP traffic ("port" 0). Figure 7.4 plots the *popularity* of the exploit ports in L_1 in the decreasing order, where the popularity of an exploit port is measured by the number of sources that have an exploit profile associated with the port. Clearly, a large majority of these ports are associated with known vulnerabilities and widely used by worms or viruses, e.g., TCP port 135 (W32/Blaster worm), TCP port 3127 (MyDoom worm). Several well-known service ports (e.g., TCP port 80, UDP port 53, TCP port 25) are also scanned/exploited by a few sources. Most sources target a single exploit, however, a small number of sources generate exploit traffic on multiple ports concurrently. In most cases, these ports are associated with the same vulnerability, for instance, the port combination {TCP port 139, TCP port 445} associated with MS Window common Internet file systems (CIFS), and {UDP ports 1026-1028} associated with MS Window messenger pop-up spams.

It is worth noting that our focus is on *significant* end hosts or services, so the sources we built behavior profiles are far less than the total number of sources seen in backbone links. Thus, it is not surprising that our behavior profiling framework identifies a subset of sources that send exploit traffic. However, these sources often account for a large percentage of exploit traffic. For example, Fig. 7.5a shows the total number of sources that send at least one flow on the most popular exploit port, port 135, as well as the number of significant sources extracted by our clustering technique that targeted port 135. As illustrated in Fig. 7.5b, the percentage of such significant sources ranges from 0% to 26%. However, as shown in Fig. 7.5c, these significant sources account for 80% traffic on TCP port 135 for most intervals. This

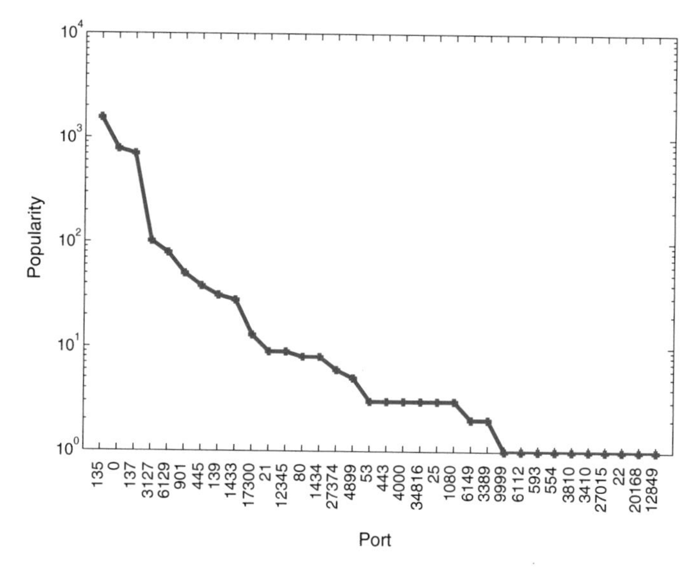

Fig. 7.4 Port popularity of exploits traffic in L_1 during a 24-h period

observation suggests that our profiling framework is effective to extract most exploit traffic sent by a small number of aggressive sources.

7.2.2 *Characteristics of Unwanted Exploit Traffic*

We study the characteristics of the exploit traffic in terms of network origins, frequency, intensity, and target footprints in the IP space. The in-depth analysis and characterization of unwanted exploit traffic provides crucial insights into exploring and developing effective strategies for reducing such traffic.

7.2.2.1 Origins of Exploit Traffic

Understanding the network origins and distributions of the exploit source addresses is very important for developing techniques of reducing unwanted traffic [13]. We first examine where the sources of exploit traffic are from, in terms of their origin ASes (autonomous systems) and geographical locations. Among the 3728 `srcIPs` in L_1 during a 24-h period, 57 are from the private RFC1918 space [14]. These source IP addresses are likely leaked from NAT boxes or spoofed. For the remaining `srcIP`'s, we search its network prefix using the *longest prefix* match in a snapshot of the BGP routing table of the same day from Route-Views [15], and obtain the AS that originates the prefix. These 3671 `srcIP`'s are from 468 different ASes.

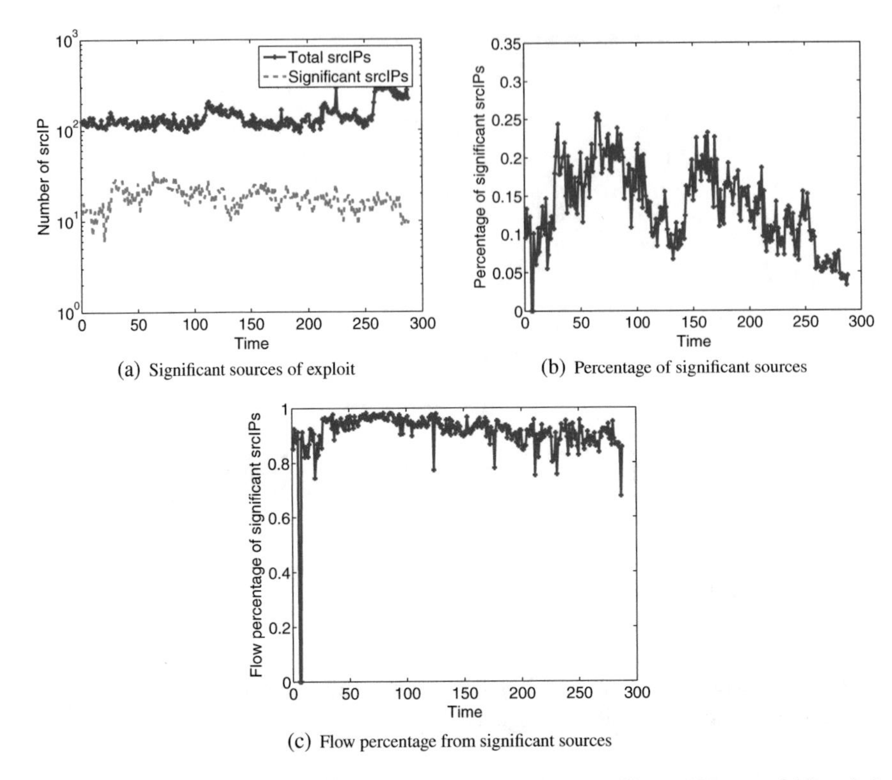

(a) Significant sources of exploit

(b) Percentage of significant sources

(c) Flow percentage from significant sources

Fig. 7.5 Aggregated traffic from significant sources of exploit on TCP port 135 over a 24-h period, i.e., 288 five-minute periods

Figure 7.6 shows the distribution of the exploit sources among these ASes. The top 10 ASes account for nearly 50% of the sources, and 9 out of them are from Asia or Europe.

7.2.2.2 Severity of Exploit Traffic

We introduce several metrics to study the temporal and spatial characteristics of exploit traffic. The *frequency*, T_f, measures the number of 5-min time periods (over the course of 24 h) in which a source is profiled by our methodology as having an exploit profile. The *persistence*, T_p, measures (in *percentage*) the number of *consecutive* 5-min periods over the total number of periods that a source sends significant amount of exploit traffic. It is only defined for sources with $T_f \geq 2$. Hence $T_p = 100(\%)$ means that the source continuously sends significant amount of exploit traffic in all the time slots it is observed. We use the *spread*, F_s, of the target footprint (i.e., destination IP address) to measure the number of /24 IP address blocks that a source touches in a 5-min time period, and the *density* of the target footprint, F_d, to

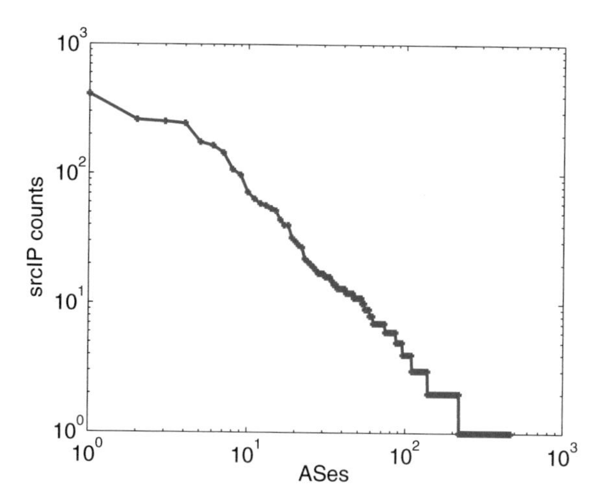

Fig. 7.6 Distribution of srcIP counts across all ASes for 3728 sources of exploit in L_1 during a 24-h period

measure the (average) number of IP addresses within each /24 block that a source touches in the period. Finally, we use the *intensity*, I, to relate both the temporal and spatial aspects of exploit traffic: it measures the (average) number of distinct target IP addresses per minute that a source touches in each 5-min period. Thus, it is an indicator how fast or aggressive a source attempts to spread the exploit.

Figure 7.7a–d show the distributions of the frequency vs. persistence, a scatter plot of the spread versus density of target footprint, the distribution of intensity, and the distributions of frequency versus intensity for the 3728 exploit sources, respectively. From Fig. 7.7a, we observe that frequency follows a power-law like distribution: only 17.2% sources have a frequency of 5 or more, while 82.8% sources have a frequency of less than 5. In particular, over 70% of them have frequency of 1 or 2. Furthermore, those 17.2% frequent ($T_f \geq 5$) sources account for 64.7%, 61.1%, and 65.5% of the total flows, packets, and bytes of exploit traffic. The persistence varies for sources with similar frequency, but nearly 60% of the sources ($T_f \geq 2$) have a persistence of 100 (%): these sources continuously send exploit traffic over time and then disappear.

From Fig. 7.7b, we see the exploit sources have quite diverse target footprints. In nearly 60% cases, exploit sources touch at least ten different /24 blocks with a density of above 20. In other words, these sources probe an average of more than 20 addresses in each block. Exploit activities with such footprint could be easily detected at the destination networks by intrusion detection systems, such as SNORT [16], and Bro [17] or portscan detecting techniques [6]. However, in about 1.6% cases, the sources have a density of less than 5, but a spread of more than 60. In a sense, these sources are smart in selecting the targets as they have a low density in each block. As the rate of exploit seen from each destination network is slow [18], they may evade port scan detection mechanisms. Upon close examination, we find that these sources employ two main strategies for target selections. One is to randomly generate targets (or to use a hit-list). The other is to choose targets like $a.b.x.d$ or $a.x.c.d$, instead of $a.b.c.x$, where x ranges from 1 to 255, and a, b, c, d take constant values.

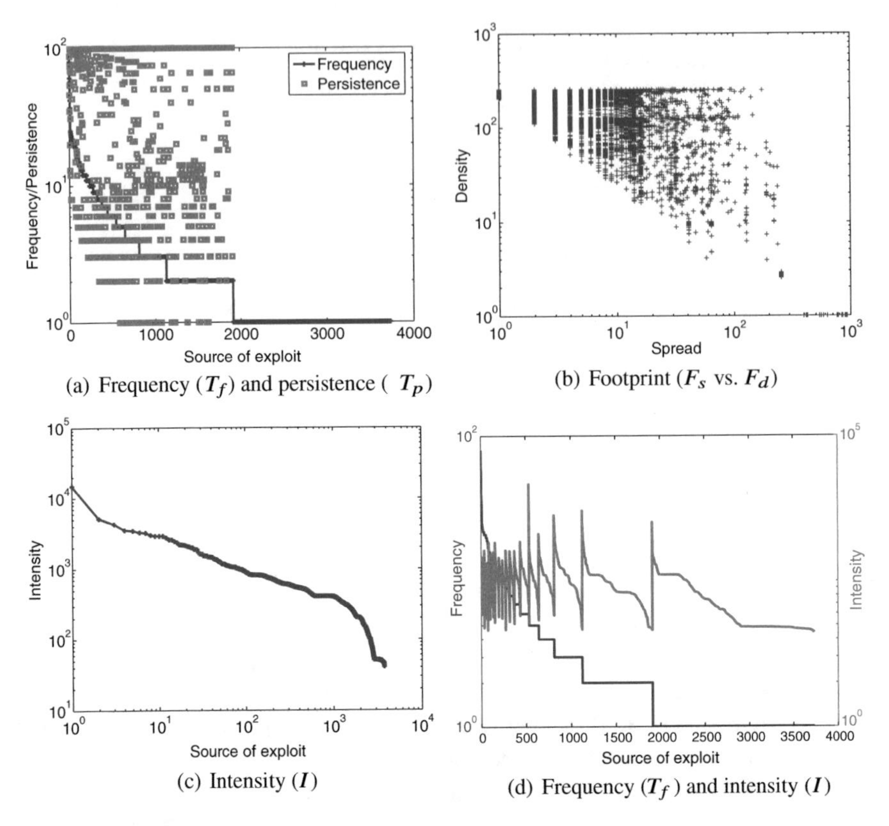

(a) Frequency (T_f) and persistence (T_p)

(b) Footprint (F_s vs. F_d)

(c) Intensity (I)

(d) Frequency (T_f) and intensity (I)

Fig. 7.7 Temporal and spatial aspects of exploit traffic for the sources with exploit profiles in the backbone link during a 24-h period. Note that **a** and **d** have the same index in x axis

The exploit intensity (Fig. 7.7c) also follows a power-law like distribution. The maximum intensity is 21K targets per minute, while the minimum is 40 targets per minute. There are only 12.9% sources with an intensity of over 500 targets per minute, while nearly 81.1% sources have an intensity of less than 500 targets per minute. Those 12.9% aggressive ($I \geq 500$) sources account for 50.5%, 53.3%, and 45.2% of the total flows, packets, and bytes of exploit traffic. However, as evident in Fig. 7.7d, there is no clear correlation between frequency and intensity of exploit traffic: the intensity of exploit activities varies across sources of similar frequency.

In summary, we see that there is a relatively small number of sources that frequently, persistently, or aggressively generate exploit traffic. They are candidates for blocking actions. Whereas a small percentage of sources are also quite smart in their exploit activities: they tend to come and go quickly, performing less intensive probing with wide-spread, low-density target footprint. These sources may be operated by malicious attackers as opposed to innocent hosts infected with malware that attempt to self-propagate. These sources need to be watched for more carefully.

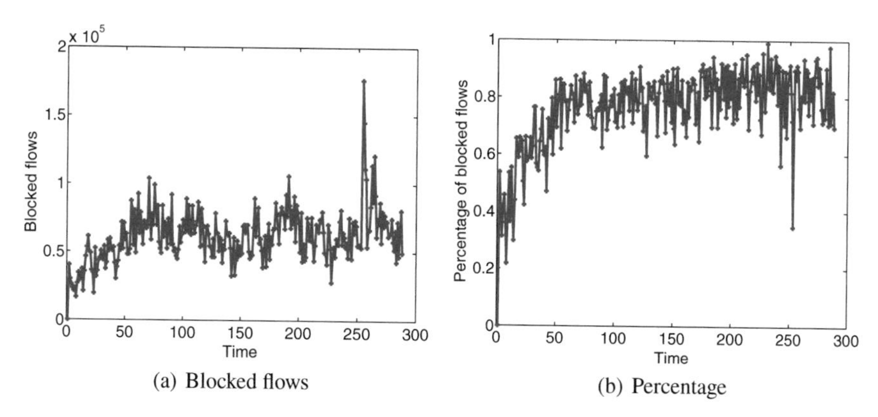

(a) Blocked flows (b) Percentage

Fig. 7.8 **a** blocked flows using the benchmark rule on L_1 over a 24-h period; **b** percentage of blocked flows over the total flows from sources of exploit

7.2.3 Strategies of Reducing Unwanted Traffic

The insights from studying the characteristics of unwanted exploit traffic allow us to design and develop several heuristic rules of blocking strategies based on characteristics of exploit activities. In order to determine which sources to block traffic from, we first use the behavior profiling technique. For every 5-minute interval, we profile all sources and identify those that exhibit the exploit traffic profile. We then devise simple rules to select some or all of these sources as candidates for blocking. Instead of blocking all traffic from the selected sources, we consider blocking traffic on only the ports that a source seeks to exploit. This is because exploit hosts may indeed be sending a mixture of legitimate and exploit traffic. For example, if an infected host behind a NAT box is sending exploit traffic, then we may observe a mixture of legitimate and exploit traffic coming from the single IP address corresponding to the NAT box.

For our evaluation, we start with the following benchmark rule. If a source is profiled as an exploit source during any 5-minute interval, then all traffic from this source on vulnerable ports is blocked from then on. Figure 7.8a, b illustrates the total blocked flows from sources of exploit every 5-min interval in L_1, and the percentage of such flows over all traffic from these sources, respectively. Overall, the benchmark rule could block about 80% traffic from the sources of exploit. In other words, this rule may still not block all traffic from the source due to two reasons. First, the source might already have been sending traffic, perhaps legitimate, prior to the time slot in which it exhibited the exploit profile. Second, as explained above, only ports on which we see exploit traffic are considered to be blocked.

While this benchmark rule is very aggressive in selecting sources for blocking, the candidate set of source/port pairs to be added to the ACLs of routers may grow to be very large across all links in a network. Therefore, we consider other blocking rules that embody additional (and more restrictive) criteria that an exploit source must satisfy in order to be selected for blocking.

- *Rule 2*: an ACL entry is created if and only if the source has been profiled with an exploit behavior on a port for n consecutive intervals. This rule is to block traffic from persistent sources;
- *Rule 3*: an ACL entry is created if and only if the source has an average intensity of at least m flows per minute. This rule is to block aggressive sources;
- *Rule 4*: an ACL entry is created if and only if the source is exploit one of the top k popular ports. This rule is to block exploit traffic of the popular ports;
- *Rule 5*: Rule 2 plus Rule 3.

We introduce three metrics, *cost*, *effectiveness*, and *wastage* to evaluate the efficacy of these rules. The cost refers to the overhead incurred in a router to store and lookup the ACLs of blocked sources/ports. For simplicity, we use the total number of sources/ports as an index of the overhead for a blocking rule. The effectiveness measures the reduction of unwanted traffic in terms of flow, packet and byte counts compared with the benchmark rule. The resource wastage refers to the number of entries in ACLs that are never used after creations.

Table 7.2 summarizes these rules of blocking strategies and their efficacy. The benchmark rule achieves the optimal performance, but has the largest cost, i.e., 3756 blocking entries. It is important to note that the cost exceeds the total number of unique sources of exploit since a few sources have exploit profiles on multiple destination ports. *Rule 2* with $n = 2$ obtains 60% reductions of the benchmark rule with 1585 ACL entries, while *Rule 2* with $n = 3$ obtains less than 40% reductions with 671 entries. *Rule 3*, with $m = 100$ or $m = 300$ achieves more than 70% reductions with 2636 or 1789 entries. *Rule 4* has a similar performance as the benchmark rule, but its cost is also very high. The *Rule 5*, a combination of *Rule 2* and *Rule 3* has a small cost, but obtains about 40% reductions compared with the benchmark rule.

We observe that the simple rules, *Rule 3* with $m = 100$ or $m = 300$ and *Rule 2* with $n = 2$, are most cost-effective when used to block the aggressive or frequent sources that send a majority of self-propagating exploit traffic, in particular, in the early stage of a malware outbreak, to hinder their spread.

7.2.4 Sequential Behavior Analysis

Next we analyze the success rate of exploit traffic and examine the follow-up activities that successful exploits engage in. We also study the communication patterns of exploit sources before they send exploit traffic to uncover potential triggers for exploit activities. The objective is to gain insights as to what kind of sources we need to watch more carefully to stop future exploit traffic.

Table 7.2 Simple blocking strategies and their efficacy

Rule	Cost	Effectiveness (Reduction (%))			Wastage
		Flow	Packet	Byte	
Benchmark	3756	–	–	–	1310
Rule 2 (n = 2)	1586	63.0%	61.2%	56.5%	505
(n = 3)	671	38.0%	36.0%	31.2%	176
Rule 3 (m = 100)	2636	97.1%	94.0%	89.4%	560
(m = 300)	1789	84.3%	80.4%	72.7%	302
(m = 500)	720	57.6%	57.0%	53.1%	68
Rule 4 (k = 5)	3471	87.4%	79.2%	77.5%	1216
(k = 10)	3624	92.9%	85.5%	81.5%	1260
Rule 5 (n = 2, m = 300)	884	48.7%	44.0%	37.7%	163

7.2.4.1 Follow-Up Activity of Exploit Traffic

We first devise a simple heuristic to infer whether a source is successful in compromising a host using TCP exploits, based solely on packet headers of one-way traffic. Since to establish a normal TCP connection, a `srcIP` and a `dstIP` must go through the three-way handshake process by exchanging SYN, SYNACK and ACK packets, with at least two packets (with SYN, and ACK bits set) from the `srcIP`, before any data packets can be exchanged. Hence for a source to successfully compromise a destination, the corresponding exploit flow must contain at least 3 packets (with the appropriate TCP flags set). Whereas TCP exploit flows with only 1 or 2 packets are clearly unsuccessful—they are either blocked, rejected or the target host does not exist. Hence we use the number of TCP exploit flows with 3 or more packets to estimate the success rate (an upper bound) of exploit traffic. In Fig. 7.9, we show the probability distribution of packet counts for the most popular (TCP135) exploit traffic. We see that more than 98% flows consist of a single packet, while over 99.5% flows have fewer than 3 packets. Hence these flows represent failed exploit attempts. This is not surprising since many networks filter incoming traffic on TCP port 135, and most patched machines do not respond to such traffic. On the other hand, about 4.2% flows have exactly 3 packets, while 676 flows (0.008%) have more than 3 packets. Similar observations hold for other TCP exploits.

In general, most exploit traffic are unsuccessful as the flows have one or two packets. On the other hand, a small fraction of flows do have a relatively large number of packets. Thus, these `srcIP`'s and `dstIP`'s are worth investigation since the dstIPs might be compromised with worms or trojans and subsequently propagate the malware.

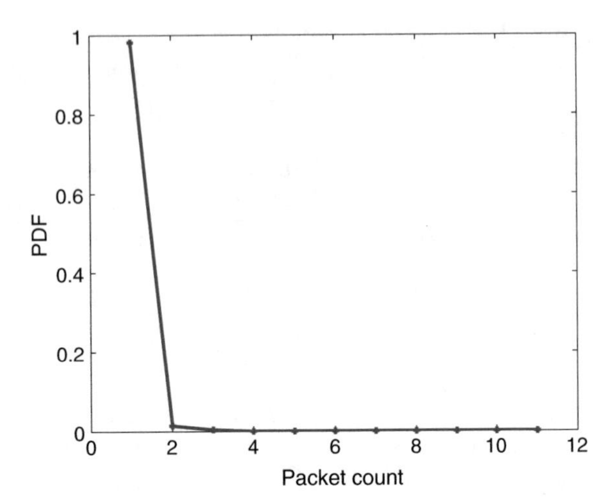

Fig. 7.9 PDF of packet counts for TCP135 exploit traffic

Intuitively, a source is very likely to take further actions if the exploit is successful. To capture what actions they typically take, we analyze their communication patterns after we observe the deviations in packet or byte counts in exploit traffic. These follow-up activities can broadly classified as follows.

- *more focused scans after probing*: the most common follow-up activity is focused scanning after probing. For example, a `srcIP` first pings a large number of end hosts to discover live hosts. Then it performs specific port scanning (e.g., TCP80, TCP135, and TCP139) on these live hosts for vulnerabilities.
- *downloading malcode to create zombies or install backdoor*: the most harmful follow-up activity is the establishment of backdoors or installation of malcode by attackers. This can be inferred by either the additional data packets after the three packets following the initial scan or subsequent communications between the attacker and the newly compromised hosts. In a number of cases, we find `srcIP`'s sends traffic to TCP4444 to a subset of previous targets which are successfully exploited on TCP135. The correlations between these ports indicate that the `srcIP` successfully infected the vulnerable hosts and setup a backdoor program to download malcode.
- *nothing at all*: For some successful exploits, we do not observe any follow-up activities in the monitoring period. However, this might simply mean that the attackers may perform further actions at a later time.

7.2.4.2 Cycles of Exploit Activities

Follow-up communication between an attacker and the compromised machines may contain the potential trigger for new exploit activities. For example, an attacker may login to a compromised machine via the backdoor, or control the zombies through IRC channels to launch new exploits,

Our method of finding such triggers is to monitor the communication patterns from the hosts likely to have been compromised. However, Internet asymmetry routing makes this very hard in the backbone network. Instead, we use packet traces collected between all hosts in Sprint cellar data network (CDN) and the Internet. These traces contain both incoming and outgoing traffic for all hosts in the CDN network in a 24-h period. We first identify the hosts with exploit profiles, and then study their communication patterns before exploit activities. There are indeed evidences that certain communications trigger new exploit activities. For example, an end host in the CDN network talks to an IRC server located in a campus network at 07:03:44 GMT on Apr. 2, 2004, and then sends exploit traffic to over 1600 hosts on TCP80 from 07:05:02 GMT in the next 5-min period. During this 5-min period, we also find the ongoing communications between this host and the IRC server. Not surprisingly, there is a significant amount of exploit traffic from this source. This event lasts about 2 h. After that, the IRC communication channel disappears, and the source no longer sends exploit traffic. Similarly, another source talks to the same IRC server, and it sends exploit traffic over 10 min. These cases suggest that these hosts in wireless networks are likely compromised, and controlled by an attacker using IRC channels.

In summary, we extract sources of exploit (thus unwanted) traffic from packet traces collected on backbone links using the behavioral traffic profiling methodology, and study the characteristics of exploit traffic from several aspects, such as network origins and severity. Based on the insights of characterizing exploit traffic, we investigate possible countermeasure strategies that a backbone ISP may pursue for reducing unwanted traffic. In addition, we propose several heuristic rules for blocking most offending sources of exploit traffic and evaluated their efficacy and performance trade-offs in reducing unwanted traffic. Our experimental results demonstrate that blocking the most offending sources is reasonably cost-effective, and can potentially stop self-propagating malware in their early stage of outburst.

7.3 Cluster-Aware Applications of Network Behavior Analysis

To demonstrate the practical benefits of end-host behavior clusters and application behavior clusters discovered via cluster-aware network behavior analysis and graphical modeling, we show how behavior clusters could be used to discover emerging applications and detect anomalous traffic behavior such as scanning activities, worms, and denial of service (DoS) attacks through synthetic traffic traces that combine IP backbone traffic and real scenarios of worm propagations and denial of service attacks. Thus, our proposed technique could become a valuable tool for network operators to gain a deep understanding of network traffic and to detect traffic anomalies.

Table 7.3 Traffic clusters of an example destination prefix

Cluster ID	Size	Flows	Patterns
1	20	422	(sip [87], spt *, dip [20], dpt 9050)
2	8	15	(sip [15], spt *, dip [8], dpt 80)
3	8	79	(sip [38], spt 80, dip [8], dpt *)
4	33	33	(sip [1], spt *, dip [33], dpt 445)

7.3.1 End-Host Network Behavior Clusters

7.3.1.1 Discovering Traffic Patterns in Network Prefixes

One major motivation of exploring behavior similarity is to gain a deep understanding of Internet traffic in backbone networks or large enterprise networks. Therefore, we first demonstrate the practical benefits of network-aware behavior clustering on discovering traffic patterns. The end-host traffic clusters discovered in each prefix reveal groups or clusters of traffic activities in the same prefixes, and understanding these patterns could be used for fine-grained traffic engineering.

End-host behavior clusters provide an improved understanding of traffic patterns in network prefixes compared with the aggregated traffic of network prefixes. For example, Table 7.3 lists four traffic clusters for one destination prefix with 69 active end-hosts during one time window. The first cluster consists of 20 destination hosts (*dip [20]*) to which 87 source hosts (*sip [87]*) talk on destination port 9050 (*dpt [9050]*) with random source ports (*spt* *), while the second cluster consists of 8 hosts to which 15 source hosts talk on destination port 80. In the third cluster, 38 source hosts talk to 8 hosts using source port 80. Finally, the last cluster consists of 33 hosts to which a single source host talks on the destination port 445 that is associated with well-known vulnerabilities. In other words, the last cluster is very likely corresponding to a scanning activity towards these hosts. If the traffic of this prefix is mixed together for analysis, it becomes very difficult to interpret and understand since there are multiple behavior patterns simultaneously occurring within the same prefix. However, by separating the traffic into different clusters, the behavior of each cluster becomes much easier for network operator to understand and take necessary actions.

7.3.1.2 Detecting Scanning Activities with Behavior Clusters of Destination Prefixes

One interesting finding on the behavior clusters of network prefixes is that many prefixes with tens of end-hosts have only a single cluster, i.e., all hosts in each of these prefixes talk with the same set of hosts. For example, Fig. 7.10a shows one case of such activity towards one prefix with 23 end-hosts in one time window. Upon

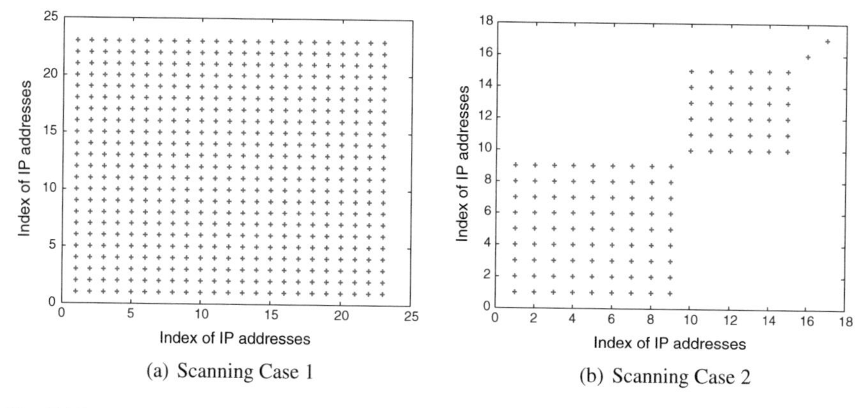

(a) Scanning Case 1 (b) Scanning Case 2

Fig. 7.10 Behavior clusters formed by scanning activities towards end-hosts in the same prefixes

close examination, we find that in this case one particular source IP scans all 23 IP addresses, thus explaining the single traffic cluster of the network prefix.

Detecting such simple scanning scenarios is not surprising, since many other existing approaches could reveal these patterns. However, the behavior clusters of destination prefixes are also able to reveal more challenging scanning cases from the massive traffic data. For instance, Fig. 7.10b shows four behavior clusters of an IP prefix. The first cluster includes nine end-hosts, while the second includes six hosts. Each of the last two clusters includes a single host since they do not share any social-behavior similarity with other hosts. By studying network traffic in each cluster, we find that the first two clusters are corresponding to two independent scanning behaviors at the same time. The first cluster is due to one scanner targeting nine different hosts, while the second cluster is caused by a different scanner targeting six other hosts. It is very interesting to note that in terms of packet counts, the last two small-sized clusters account for 99.76% of network traffic (6655 out of 6671 data packets), while the first two clusters, having only nine and seven packets respectively, accounting for a very small percentage of the traffic. If traffic analysis is simply focused on the entire prefix, such low-volume anomalous patterns could simply be missed. Therefore, this suggests that behavior analysis on host communication patterns is complementary to existing volume-based techniques for detecting scanning behavior patterns.

7.3.1.3 Detecting Worm Behavior in Its Early Phases

To demonstrate practical benefits of network-aware behavior clustering in detecting worm behavior, we use real traces of Witty worm collected by CAIDA [19] and combine it with the backbone network traffic into synthetic traffic. The behavior clustering is able to detect a new cluster in one of the prefixes during the very beginning of worm propagations. Figure 7.11a, b show behavior clusters of this prefix before

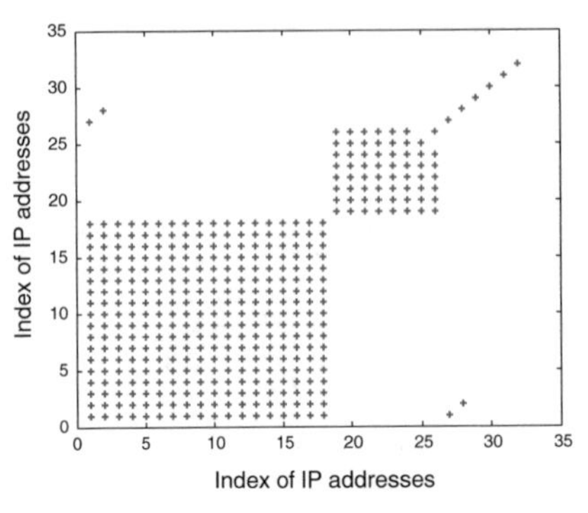

(a) Behavior clusters of a network prefix before Witty worm propagations

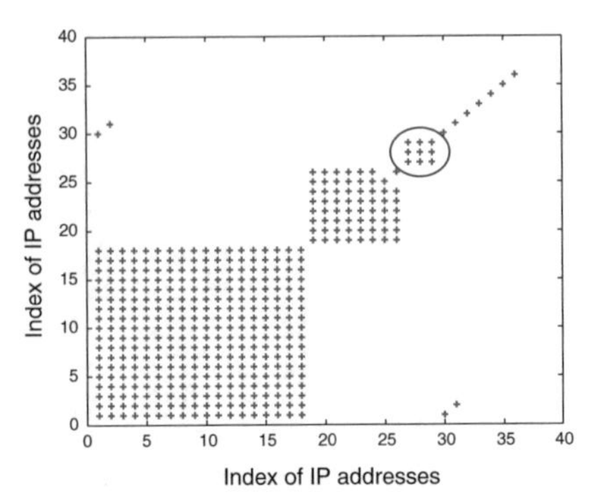

(b) Behavior clusters of the prefix during the first few minute of Witty worm propagations

Fig. 7.11 Emerging behavior clusters formed by worm propagations

and after worm propagations, respectively. An emerging small cluster consisting of three end-hosts marked by the circle in Fig. 7.11b is actually triggered by data packets containing the Witty worms. Such emerging behavior clusters of a network prefix triggered by worm propagation events or other suspicious activities serve as strong alarm signals to network operators for immediate response and in-depth analysis.

7.3.1.4 Detecting DDoS Attacks

Detecting and mitigating DDoS attacks is one of the challenging tasks facing network operators or security analysts at edge networks due to the nature of these attacks in saturating network links. However, we argue that pushing the detection from edge networks to backbone networks is beneficial, since backbone networks have sufficient bandwidth and diverse routing paths compared with edge networks. By combining backbone traffic from a large ISP and real cases of DDoS attacks identified in the previous work [20], we demonstrate the usage of behavior similarity in detecting DDoS attacks in Internet backbone networks.

Figure 7.12a–d illustrate the behavior clusters of two source IP prefixes before and during DDoS attacks based on the synthetic traffic traces. The spectral clustering reveals emerging clusters or cluster changes during DDoS attacks for both source prefixes (Fig. 7.12b, d). For the first prefix, 39 end-hosts form an emerging cluster in Fig. 7.12b, while in Fig. 7.12d the existing cluster of 25 end-hosts of the second prefix in Fig. 7.12c is expanded to a much larger cluster with 52 end-hosts. The reason for the abnormal expansion of the cluster in the second prefix is that the existing 25 hosts join other 27 hosts in the same prefix in launching the DDoS attacks while sending normal data traffic as well. Compared with other methods of detecting DDoS attacks, the advantage of behavior clusters is to leverage the small emerging clusters and the dynamics of existing clusters for capturing interesting events, such that the attacks

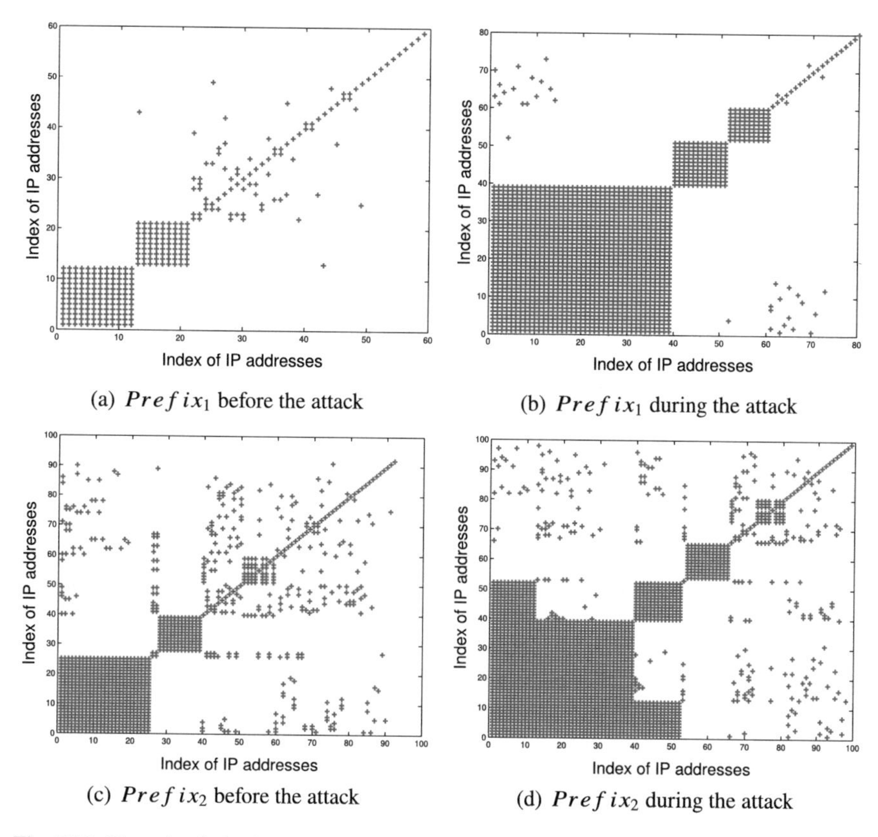

(a) $Prefix_1$ before the attack

(b) $Prefix_1$ during the attack

(c) $Prefix_2$ before the attack

(d) $Prefix_2$ during the attack

Fig. 7.12 Emerging behavior clusters of two independent source prefixes formed during DDoS attacks

could be detected before the traffic arrives at edge networks and saturates network links connecting edge networks to the Internet.

7.3.2 Network Application Behavior Clusters

To demonstrate the usage of application behavior clusters, we use case studies to illustrate how these clusters could aid in identifying emerging applications and detecting anomalous traffic patterns.

7.3.2.1 Detecting Emerging Applications

As the Internet continues to grow in end users, mobile devices, and applications, classifying Internet applications becomes more complicated due to the rapid growth of new applications, mixed uses of application ports, traffic hiding using well-known ports for avoiding firewall filtering. On the other hand, detecting emerging applications is very important for traffic engineering and security monitoring. In our experiment results, we find that application clusters are a feasible approach of finding new applications that share similar clustering coefficient or similar social behaviors of end-hosts with existing known applications. For example, we find that an unknown port consistently follows in the same clusters with service ports TCP port 25 (SMTP), TCP port 80 (Web), TCP port 443 (HTTPS). We conjecture that this port very likely corresponds to a service application since the source and destination hosts engaging in this port exhibit similar social behaviors with existing known applications. Such findings could provide very valuable information for network operators for in-depth analysis.

7.3.2.2 Detecting Anomalous Traffic Patterns

The usage of application behavior clusters on Internet traffic also includes detecting anomalous traffic patterns. For example, the clustering results of application traffic shows a cluster of TCP destination ports 135, 1433, and 22. The first two are ports associated with well-known vulnerabilities, thus it is not surprising to observe these two ports in the same cluster. However, port 22 is mostly used for SSH traffic, and it is expected to be grouped into clusters that include other major Internet service ports. An in-depth analysis reveal that during that particular time window, six source IP addresses in the same `/26` network prefix send only TCP SYN packet to 28 unique destination address on destination TCP port 22. These hosts likely scan SSH ports on Internet hosts. The scanning traffic together account for 66% of total flows towards destination TCP port 22 during the time window, which explains why TCP port 22 is clustered with ports associated with well-known vulnerabilities.

In summary, based on synthetic network traffic combining backbone network traffic and real scenarios of worm propagations and denial of service attacks, our experiments have demonstrated that behavior clusters of end hosts in the same networks are able to aid in discovering traffic patterns for traffic engineering or access control list constructions, and in detecting anomalous behavior such as scanning activities, worms, and denial of service (DoS) attacks through synthetic traffic traces. Similarly our experimental results also show that application behavior clusters are able to group Internet ports into distinct clusters based on clustering coefficient and other graph properties of source and destination behavior graphs. These behavior clusters could aid network operators in understanding emerging applications and detecting anomalous traffic towards Internet applications.

7.4 Summary

A wide spectrum of applications have benefited from the capabilities of network behavior analysis in understanding behavioral patterns of networked systems and Internet applications. These applications leverage the end-to-end process of network behavior analysis which includes the collection, storage, processing, exploration, analysis, visualization, and interpretation of network traffic data. The insights provided by data-driven network behavior analysis allows network operations and security analysts to gain an in-depth understanding of normal data communications between networked systems for various Internet applications or services, and to identify anomalous and intrusion activities towards these networked systems and Internet applications.

References

1. J. Otto, M. Sánchez, D. Choffnes, F. Bustamante, G. Siganos, On blind mice and the elephant – understanding the network impact of a large distributed system, in *Proceedings of ACM SIGCOMM* (2011)
2. R. Basat, G. Einziger, R. Friedman, M. Luizelli, E. Waisbard, Constant time updates in hierarchical heavy hitters, in *Proceedings of ACM SIGCOMM* (2017)
3. K. Xu, Z.-L. Zhang, S. Bhattacharyya, Reducing unwanted traffic in a backbone network, in *Proceedings of Steps to Reducing Unwanted Traffic on the Internet Workshop (SRUTI)* (2005)
4. R. Pang, V. Yegneswaran, P. Barford, V. Paxson, L. Peterson, Characteristics of internet background radiation, in *Proceedings of ACM Internet Measurement Conference* (2004)
5. V. Yegneswaran , P. Barford, J. Ullrich, Internet intrusions: global characteristics and prevalence, in *Proceedings of ACM SIGMETRICS* (2003)
6. J. Jung, V. Paxson, A. Berger, H. Balakrishna, Fast portscan detection using sequential hypothesis testing, in *Proceedings of IEEE Symposium on Security and Privacy* (2004)
7. T. Karagiannis, K. Papagiannaki, M. Faloutsos, BLINC: multilevel traffic classification in the dark, in *Proceedings of ACM SIGCOMM* (2005)
8. A. Ramachandran, N. Feamster, Understanding the network-level behavior of spammers, in *Proceedings of ACM SIGCOMM* (2006)
9. M. Bailey, E. Cooke, F. Jahanian, J. Nazario, D. Watson, The internet motion sensor: a distributed blackhole monitoring system, in *Proceedings of Network and Distributed System Security Symposium* (2005)
10. S. Schechter, J. Jung, A. Berger, Fast detection of scanning worm infections, in *Proceedings of Symposium on RAID* (2004)
11. V. Yegneswaran, P. Barford, V. Paxson, Using honeynets for internet situational awareness, in *Proceedings of the ACM/USENIX Hotnets IV* (2005)
12. D. Moore, V. Paxson, S. Savage, C. Shannon, S. Staniford, N. Weaver, Inside the Slammer Worm. IEEE Security and Privacy (2003)
13. P. Barford, R. Nowak, R. Willett, V. Yegneswaran, Toward a model for sources of internet background radiation, in *Proceedings of the Passive and Active Measurement Conference* (2006)
14. Y. Rekhter, B. Moskowitz, D. Karrenberg, G. J. de Groot, E. Lear: RFC1918: Address Allocation for Private Internets (1996)
15. of Oregon, U.: Routeviews archive project, http://archive.routeviews.org/
16. Snort: Snort - Network Intrusion Detection & Prevention System, http://www.snort.org/
17. V. Paxson, Bro: a system for detecting network intruders in real-time. Comput. Netw. **31**, 2435–2463 (1999)

18. S. Staniford, J. Hoagland, J. McAlerney, Practical automated detection of stealthy portscans. J. Comput. Secur. **10**, 105–136 (2002)
19. C. Shannon, D. Moore, The spread of the witty worm. IEEE Secur. Priv. **2**(4), 46–50 (2004)
20. A. Hussain, J. Heidemann, C. Papadopoulos, A framework for classifying denial of service attacks, in *Proceedings of ACM SIGCOMM* (2003)

Chapter 8
Research Frontiers of Network Behavior Analysis

Abstract As the Internet continues to bring innovative applications and services to the broad society, making sense of behavioral objects on the Internet with network behavior analysis will remain an important technique for understanding and characterizing novel network environments, emerging network applications, and new networked systems. This chapter presents the research frontiers of network behavior analysis in cloud computing, smart home networks, and the Internet of Things (IoT) paradigms. This chapter first discusses network behavior analysis as a service (NBA-as-a-service) in cloud computing environments for monitoring and securing large-scale Internet data centers. Subsequently, this chapter presents how network behavior analysis provides new traffic and behavior insights into Internet-connected devices in distributed smart home networks. Finally this chapter introduces a multidimensional network behavior analysis framework to characterize behavioral patterns of heterogeneous IoT devices in edge networks.

8.1 Network Behavior Analysis in the Cloud

Cloud computing integrates data, applications, users, and servers on a vast scale and enables a global optimization of computing resources. However, due to security threats from both outside and inside the cloud, security remains as a significant challenge and obstacle in the wide adoptions of cloud computing paradigms. To enhance the security of networks, applications, and data in the cloud, the study [1] envisions a profiling-as-a-service architecture to characterize, understand and profile network traffic at multiple layers in the multi-tenant cloud computing environment: network routers, hypervisors, virtual instances, and applications. The proposed architecture will not only provide an in-depth understanding on traffic patterns of cloud tenants, but also enhance the security of cloud computing by collaboratively detecting and filtering unwanted traffic towards cloud instances.

K. Xu, *Network Behavior Analysis*,
https://doi.org/10.1007/978-981-16-8325-1_8

8.1.1 Background

Cloud computing integrates data, applications, users, and servers on a vast scale and enables a global optimization of computing resources. However, due to security threats from both outside and inside the cloud, security remains as a significant challenge and obstacle in the wide adoptions of cloud computing paradigms [2, 3]. For example, the in cloud experiments in [4] demonstrate the vulnerabilities associated with shared virtual machines (VM) on the same physical host and the possibility of mounting cross-VM side-channel attacks to collect information from the target VMs.

In light of the potential attacks and threats towards cloud computing, security has become one of the major concerns for the adoptions of cloud computing [5]. The recent work [4] introduces the vulnerabilities with shared virtual machines (VM) from cloud computing providers and demonstrates the feasibility of mounting cross-VM side-channel attacks to gain information from the target VMs. In [6] Ertaul et al. survey security challenges in cloud computing environment, while Chen et al. identified two new facets to cloud computing security [7], namely, "the complexities of multi-party trust considerations" and "the ensuring need for mutual auditability". Yildiz et al. [8] first identifies security concerns on multiple layers arising in cloud computing, and subsequently outlines a policy-based security approach for cloud computing through defining security polices at various layers including networking, storage, systems management, and applications.

Given the magnitude and diversity of security threats towards cloud computing, it is crucial to develop effective solutions to ensure the security and high-availability of data, applications, and networks for cloud tenants. The central challenges of enhancing the security of cloud computing are (i) the vast amount of network traffic in the cloud and the diversity of cloud tenants, (ii) the variety of security threats that include traditional threats towards cloud tenants and emerging threats brought by the cloud computing paradigm, (iii) the launching points of the attacks from inside and outside the cloud; and (iv) the "untrusted" nature of the multi-tenant cloud computing environment [9]. Many recent research have been conducted on new data center architecture [10–12] and network traffic measurement in cloud computing [13]. However, there has been little attempt to profile network traffic of cloud instances. Existing techniques for cloud computing security such as access control lists or firewalls are widely deployed on data center routers and virtualization servers, however they are insufficient for securing cloud instances as cloud computing tenants face a variety of security challenges such as intrusion attempts, port scanning, and denial-of-service attacks from outside the cloud as well as from inside the cloud, e.g., cloud providers or other cloud tenants.

8.1.2 *Profiling-as-a-Service in the Cloud*

Traffic profiling has recently become an essential technique for securing and managing backbone and edge networks, e.g., building normal and anomalous network behavior profiles [14, 15], detecting traffic obfuscation and encryption [16], and accurate identification of network applications [17, 18]. The proposed *profiling-as-a-service* architecture [1] for network behavior analysis in the cloud analyzes and characterizes network traffic of cloud instances at multiple layers in the multi-tenant cloud computing environment: (1) border routers of cloud networks, (2) hypervisors of virtualization servers, (3) virtual instances (VMs), and (4) applications.

The layered *profiling-as-a-service* approach builds hierarchical traffic profiles for cloud instances and provides an in-depth understanding of network traffic towards cloud instances. The architecture consists of four system components that build upon each other to establish *profiling-as-a-service* in the cloud: (i) a layered approach of profiling network traffic of cloud instances, (ii) behavior models and structural models based on communication patterns of cloud instances, (iii) a collaborative solution for detecting unwanted traffic in the cloud, and (iv) a profiling-aware sampling algorithm for improving the robustness of the proposed architecture during sudden traffic surges caused by anomalous events. The goal of the profiling service is to provide an in-depth understanding of traffic patterns for cloud tenants and to enhance the security of cloud computing by collaboratively detecting and filtering unwanted traffic towards cloud instances.

Network traffic profiling has been extensively studied in the recent years for understanding network traffic in Internet backbone networks and edge networks [14, 15, 17]. For example, [15] builds behavior profiles of end hosts and network applications using traffic communication patterns without any presumption on what is normal or anomalous, while in [17] the authors study the host behaviors at three levels with the objective to classify traffic flows using packet header information only. Jiang et al. [14] creates a traffic profile for each network prefix through behavior analysis of aggregated traffic. Different from these work, the *profiling-as-a-service* architecture for the cloud attempts to build traffic profiles across all layers in the multi-tenant cloud computing environments for improved security and management in the cloud.

8.1.3 *Architecture of Profiling-as-a-Service for Network Behavior Analysis*

In light of the security and privacy challenges of cloud computing, it becomes increasingly important for cloud customers to know *what happens to their cloud instances* in the multi-tenant cloud computing environment managed by a third-party cloud provider. Towards this end, we build the *profiling-as-a-service* architecture for establishing hierarchical traffic behavior profiles of cloud instances at multiple layers—border routers of cloud networks (network profiling), hypervisors of virtualization

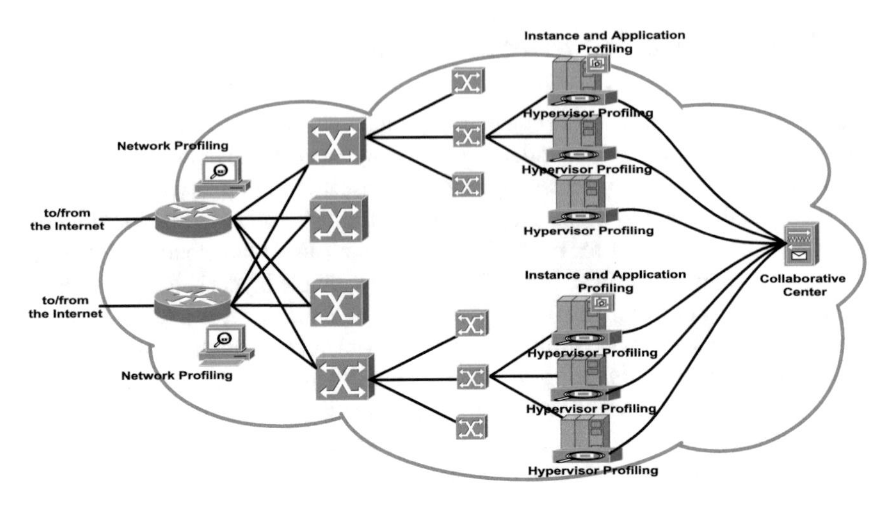

Fig. 8.1 Architecture of layered profiling in cloud computing

servers (hypervisor profiling), virtual instances (instance profiling), and applications (application profiling). Figure 8.1 illustrates a schematic diagram of the layered *profiling-as-a-service* architecture, in which each layer analyzes cloud traffic from a different perspective and thus provides a unique insight on traffic patterns of cloud instances.

Figure 8.2 shows the hierarchical relationships of these four levels in the cloud hierarchy—network-level, hypervisor level, instance level, and application level. The lower two layers, network and hypervisor profiling, analyze network flows and unwanted traffic of all cloud instances at the routers and hypervisors with coarse granularity, while the higher two layers focus on the fine-grained traffic analysis for individual instances or their applications. Thus, in the proposed *profiling-as-a-service* architecture, each layer provides a unique view to traffic profiles and behavior patterns of cloud instances.

The major intuition of the layered profiling approach lies in that each layer of the cloud computing environment provides a unique perspective on network traffic of cloud instances. For example, in the experiments of mounting cross-VM side-channel attacks [4] the step of network probing for building cloud cartography utilizes wget scan to determine the liveness of EC2 instances, and such activities would leave distinct traffic footprints at each level of the cloud hierarchy and lead to different traffic profiles at four levels. Hence, the combined insights from each layer lead to an in-depth understanding on the traffic patterns of cloud instances. We could in turn use the complementary profiles in these layers to build a comprehensive and correlated traffic profile of cloud instances. The practical applications of the first two layers include (i) understanding network-level traffic patterns of cloud instances and (ii) correlating cloud-wide unwanted traffic towards cloud instances, while the applications of the latter two layers include (i) revealing the overall traffic patterns and

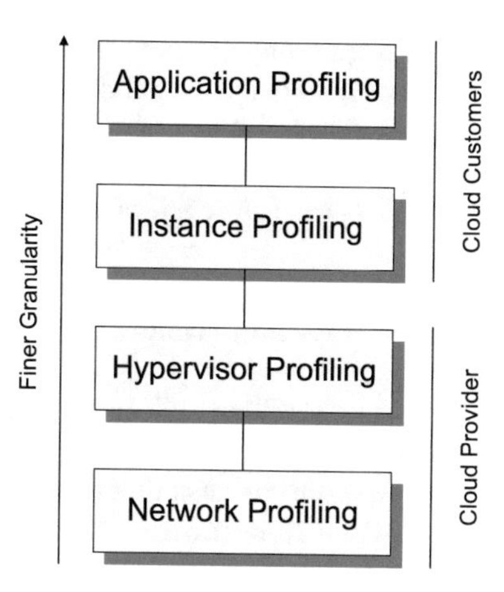

Fig. 8.2 Hierarchical relationships of the profiling layers

end-user access behaviors of cloud instances and (ii) generating application-specific traffic patterns or detecting malicious packet payload to cloud applications.

8.1.4 *Designing the Profiling-as-a-Service Infrastructure*

In the rest of this section we will describe how to profile network traffic at each layer and address the inherent challenges of the *profiling-as-a-service* architecture, such as large volume of network traffic during denial-of-service attacks. Specifically, we will address these following problems: (i) how to build the *profiling-as-a-service* architecture in the cloud networks, (ii) what traffic features should be included in traffic profiles, (iii) how to profile network traffic and behavior patterns in normal conditions and anomalous events, and (iv) how to correlate profiles from distributed hypervisors and instances in the cloud for collaborative security monitoring.

8.1.4.1 Profiling Traffic at Network-Level

The first step of the *profiling-as-a-service* infrastructure focuses on the incoming or outgoing traffic observed at the border routers that connect cloud networks to the Internet. Profiling traffic at the network-level provides a broad view of traffic patterns of cloud instances. However, due to the sheer volume of network traffic to/from thousands of cloud instances, it remains a daunting task to analyze vast network traffic in the cloud networks. Therefore, the size of cloud traffic data calls for lightweight and efficient algorithms to make sense of these traffic and to generate meaningful traffic summaries of cloud instances at the cloud network level. Towards this end, we plan to explore entropy concepts from information theory and histogram analysis to analyze the distribution of traffic features for cloud instances at the network level.

Network profiling analyzes the incoming and outgoing traffic of border routers at the cloud networks, and builds high-level traffic summaries of all instances through network flows. These profiles can be used to correlate common threats, such as worms and network-wide scanning, and to detect and mitigate volume-based traffic anomalies, such as denial-of-service attacks.

8.1.4.2 Profiling Traffic at Hypervisor Level

Hypervisor profiling collects, analyzes, and correlates "unwanted traffic" captured by firewalls deployed on the hypervisor layer of virtualization servers that support multiple instances. In addition, we establish a central collaborative center that communicates with distributed hypervisors in the cloud, correlates "unwanted traffic" filtered by distributed hypervisor firewalls, and reports the aggregated trends of security threats to all cloud instances. Hence, hypervisor profiling could become a powerful technique for detecting low-volume attacks such as scanning activities and penetration attempts towards cloud instances.

One of the key technologies that drive the cloud computing paradigm is the use of virtualization, which allows multiple instances (also called virtual machines) to run on the same physical machine. These multiple instances are isolated from each other through the hypervisor layer (also called virtualization layer). The hypervisor layer arbitrates and manages CPU, physical memory, and I/O devices among multiple instances running on the same machine. All data packets from or to cloud instance pass through the hypervisor layer, as this layer resides between the physical network interface and the virtual network interface of cloud instances. Therefore, the host-based firewall is often deployed at this layer to filter "unwanted traffic" towards the cloud instances using pre-defined firewall policy rules configured by cloud customers. For example, Amazon EC2 allows cloud customers to configure security polices to define certain firewall rules at the hypervisor layer for accurately identifying and filtering the inbound "unwanted traffic" to the cloud instances [19].

In this study, we propose to harvest unwanted traffic from distributed hypervisors in the cloud and to establish a central collaborative center that collects and analyzes unwanted traffic from distributed hypervisors in the cloud computing environment. The idea of this collaborative center is inspired by DShield, a cooperative network security community portal site that collects firewall logs for analyzing the trends and emerging threats of the exploit behaviors [20]. Specifically, we will develop a distributed measurement framework, where each hypervisor has a client program that communicates with a server running on the collaborative center. Once a hypervisor detects unwanted traffic towards one or more instances, the hypervisor will summarize and generate the traffic signature, and then sent to the central collaborative center through the reporting client. Prior studies have shown that the collaborative principles have a wide range of applications in network measurement [21] and security monitoring [22].

Given the volume of unwanted traffic in the Internet background radiations due to vulnerability scanning, worm propagations, system penetration attempts, DoS attacks, and other exploit activities [23], it is not surprising that cloud instances receive a large amount of unwanted traffic. Thus a major challenge of designing and implementing such a collaborative solution for distributed hypervisors in the cloud computing environment lies in the size and diversity of the "unwanted traffic" collected and processed by distributed hypervisors and the central collaborative center. Thus an important research problem in hypervisor profiling is how to develop efficient collaborative algorithms to share and leverage unwanted traffic collected on distributed hypervisors for reducing unwanted traffic towards cloud instances. To prevent distributed hypervisors and the collaborative center from being overwhelmed by a large amount of unwanted traffic, we propose to use two-layer counting bloom filter technique [24] at hypervisors and the collaborative center to reduce the size of unwanted traffic reports by identifying the most aggressive attackers from all source IP addresses of unwanted traffic.

8.1.4.3 Profiling Traffic at Instance Level and Application Level

Network and hypervisor profiling in the proposed *profiling-as-a-service* architecture discover behaviors patterns of cloud instances and unwanted traffic towards the cloud network, respectively. However, both steps lack the visibility of all network traffic towards cloud instances and the applications running on them. To gain a complete picture of network traffic towards cloud instances and their applications, it becomes very necessary to profiling traffic at instance and application levels.

Instance profiling is interested in three important aspects of cloud instances: user, traffic, and performance. The user profiling is focused on the access patterns of end systems on the Internet that communicate with cloud instances, while traffic profiling at the instance-level studies traffic characteristics of the cloud instance, such as traffic distributions across ports or applications, and temporal traffic patterns. Similar to the continuous profiling infrastructure deployed at Google data centers [25], the performance profiling aims in quantifying system performance such as CPU and memory utilization, and input/output throughput, and end-to-end performance such as network latency and packet losses. As illustrated in Fig. 8.3, instance profiling runs independently on multiple virtual instances that are hosted on the same physical machine.

In addition to profiling traffic at instance level that studies the aggregated network traffic of cloud instances, we will also perform profiling traffic at application level and investigate the application-specific semantics and contents of network traffic. We plan to use the graphic models to represent traffic activity of cloud instances and their applications. The fine-grained traffic analysis on application level provides valuable insights for application diagnosis and troubleshooting, network management and capacity planning.

The challenges of instance profiling and application profiling are (i) feature selection for instance or application profiling; (2) payload and content analysis without baseline signatures or prior knowledge of normal or abnormal traffic patterns. To address these challenges, we will employ temporal analysis and feature selection

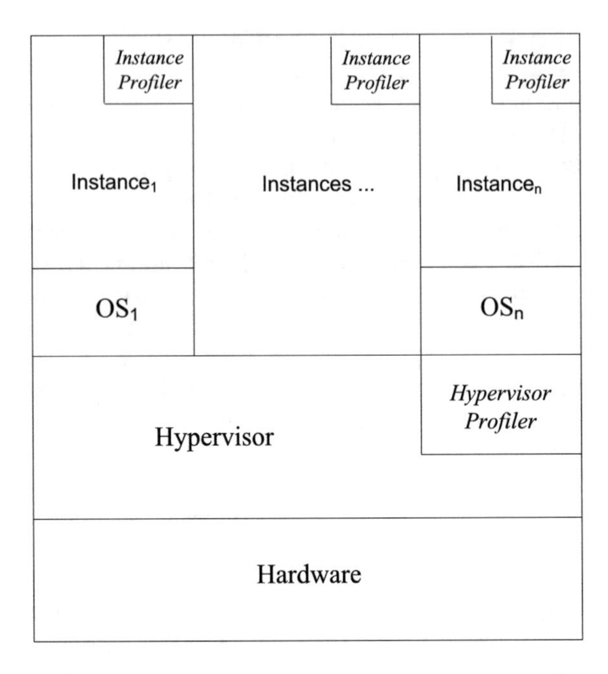

Fig. 8.3 Hypervisor profiling and instance profiling on the physical machine

algorithms for instance and application profiling, and explore algorithms in data mining and machine learning for detecting unknown exploit traffic, e.g., emerging worms or virus. The ultimate goal of profiling traffic at instance level and application level is to complement traffic profiles of network profiling and hypervisor profiling and to build a comprehensive traffic behavior profile for each cloud instance by summarizing its traffic behavior and application activities with the full packet traces.

8.1.4.4 Profiling High-Volume Network Traffic

An operational challenge of the *profiling-as-a-service* architecture across all levels is the sudden traffic surges during unusual events such as denial-of-service attacks [26], flash crowds [27], or worm outbreaks [28]. The sheer volume of network traffic during these events introduces significant system pressure for the profiling architecture that runs on commodity PCs with limited CPU and memory capacity. At the same time, it is vital for the profiling architecture to function during these events, since traffic profiles generated during these periods will provide key insights and valuable information for effective response and forensic analysis.

Sampling is a widely deployed technique to reduce system resource consumptions in network traffic monitoring. Traditional sampling approaches include random packet sampling, random flow sampling, smart sampling and sample-and-hold algorithms. However, previous studies [29] have shown that these existing sampling algorithms, albeit significantly reducing resource usage, bring non-trivial accuracy losses on the traffic feature distributions as well as on other traffic statistics. Our preliminary analysis also finds that these sampling algorithms could lead to inaccurate traffic profiles of cloud instances at all levels during sudden traffic surges, although they substantially bring down the CPU and memory usage. To enhance the robustness and accuracy of the profiling-as-a-service architecture under these stress conditions, it becomes very necessary to develop novel sampling algorithms that not only reduce the system resource usage, but also retain the accuracy of traffic profiles across all levels.

A key lesson from studying traditional sampling approaches in our preliminary analysis is that the cloud instances involved with anomalous events such as denial-of-service attacks usually receive vast amount of network traffic. Profiling traffic behavior of these instances during these events does not require a very large number of traffic flows, since their feature distributions likely remain the same even with a small percentage of sampled traffic flows. On the other hand, the profiles of other hosts with much less traffic are very sensitive to the number of sampled traffic. Based on this insight, we plan to develop new profiling-aware sampling solutions that limit the number of sampled traffic flows for instances or applications with a large amount of traffic, but adaptively samples on the rest of instances or applications when the profiling system is faced with sudden explosive growth in the number of traffic flows or packets caused by anomalous events such as denial-of-service attacks or worm outbreaks.

The success and challenges of cloud computing have recently draw broad attentions from the networking and system research community. In a view of cloud computing [3], Armbrust et al. summarize top 10 obstacles and research opportunities for cloud computing. As a first step of understanding network traffic in the cloud, the *profiling-as-a-service* infrastructure in the multi-tenant cloud computing environment builds traffic profiles of cloud instances at multiple layers for providing critical insights on network behavior analysis, security monitoring, and traffic engineering for cloud instances. To demonstrate the operational feasibility and the practical applications of the *profiling-as-a-service* architecture, the next important research task is to design, implement, and evaluate a prototype profiling system in real Internet data centers and existing commercial cloud computing platforms such as Amazon EC2, Microsoft Azune, or Google AppEngine.

8.2 Network Behavior Analysis in Smart Homes

The rapid spread of residential broadband connections and Internet-capable consumer devices in home networks has changed the landscape of Internet traffic. To gain a deep understanding of Internet traffic for home networks, this section presents a traffic monitoring platform that collects and analyzes home network traffic via programmable home routers and traffic profiling servers. Using traffic data captured from real home networks, we present traffic characteristics in home networks, and then apply principal component analysis to uncover temporal correlations among application ports. In light of prevalent unwanted traffic on the Internet, we characterize the intensity, sources, and port diversity of unwanted traffic towards home networks.

8.2.1 Background

In recent years, the rapid growth of Internet-capable devices in the home and residential broadband access has driven the rising adoptions of home networks. The availability of home networks not only creates new application opportunities such as remote health care and Internet television, but also changes the distribution of Internet traffic, e.g., a recent study shows that video streaming via Netflix accounts for 32.7% of peak downstream traffic in United States [30]. As home networks become an important part of the Internet ecosystem, it is very crucial to understand network traffic between the Internet and home networks as well as the traffic exchanged within home devices.

Most home users lack technical expertise to manage the increasingly complicated home networks, and an extensive body of research have focused on how to simplify network management tasks for home users [31–35]. Several recent studies have been devoted to understanding traffic characteristics of home networks using aggregated and sampled traffic collected from edge routers in Internet service providers [36–38]. However, these measurement studies stand from the perspective of outside home networks, thus lack the visibility of *what is happening in home networks*. The in-depth understanding of home network traffic could aid home users in effectively securing and managing home networks.

Unlike enterprise networks which have dedicated network professionals to manage and operate the networks, securing home networks have been a considerable challenge, as most home users do not have sufficient technical expertise and knowledge to manage and secure the networks [39]. As a result, connected devices in home networks are targets and victims of virus, worms, and botnets, and become a major source of spams and a part of botnets. In [40], Feamster proposes to outsource the management and operations of home networks to a third party that has expertise of network operations and security management. In [41], Yang et al. study network management tools that are currently deployed in home networks via interviewing 25 home networks users, and report user experiences of these network management tools. To aid in troubleshooting and managing home networks, [42] proposes to build a home network data recorder system as a general-purpose logging platform to record what is happening in home networks. Many researches have also focused human-computer interactions in home networks [33–35], troubleshooting and diagnosis [31, 32], and broadband network sharing among different Internet service providers [43].

Home network performance has recently drawn significant attentions from the research community. A recent work [44] performs controlled experiments in a lab environment for evaluating the impact of home networks on end-to-end performance of end systems. In addition, several commercial or open-source tools have been developed for measuring and diagnosing Internet properties of end users. For example, Netalyze [45], a network measurement and diagnosis service, tests a wide variety of functionalities at network, transport and application layers for end users' Internet connectivity in edge networks such as home networks. Kermit, a network probing tool, was developed in [46] to visualize the broadband speed and bandwidth usage for home users. Hatonen et al. [47] measures and analyzes the behavior characteristics of a variety of home gateways such as DSL and cable modems, including NAT binding timeout, throughput, and protocol support, and their influence on network performance and user experience. A recent work [48] measures network access link performance directly from home gateway devices, and has inspired us to characterize network traffic from inside home networks through programmable home routers.

As residential broadband users continue to grow, many studies have been devoted to measure and characterize residential broadband networks [36–38]. However, all of these studies stand from outside home networks, and lack the visibility of the home networks, such as home network architecture, diversity of end hosts. For example, [37] examines the growth of residential user-to-user traffic in Japan, a country with a high penetration rate of residential broadband access, and studies the impact of these traffic on usage patterns and traffic engineering of commercial backbone networks. In addition, [38] studies several properties of broadband networks, including link capacities, round-trip times, jitter, and packet loss rates using active TCP and ICMP probes, while [36] passively collects packet-level traffic data of residential networks at aggregated routers of a large Internet service provider, and analyzes dominant characteristics of residential traffic including network and transport-level features, prominent applications, and network path dynamics. Different from these prior work, the study of network behavior analysis in smart homes [49] leverages the availability of traffic flows exported from programmable home routers, and presents the first study of traffic characteristics of Internet-capable devices in home networks.

8.2.2 Traffic Monitoring Platform for Home Networks

In light of the rapid growth of home networks, we have presented a traffic monitoring platform to collect and analyze network traffic for Internet-capable devices in home networks. Relying on programmable home routers that connect home networks to the Internet, the platform collects network flow streams to traffic profiling servers, and analyzes traffic characteristics of home networks. Our findings on traffic characteristics of application ports lead us to explore principal component analysis (PCA) to uncover temporal correlations among these ports. The experiment results show that there indeed exist several application port clusters in home networks with each cluster exhibiting distinct traffic patterns. Given the new perspective of understanding unwanted traffic on the Internet from network behavior analysis in home networks, the next two important research tasks are to develop privacy-preserving data collection capacity into the traffic monitoring platform and to deploy the platform into a

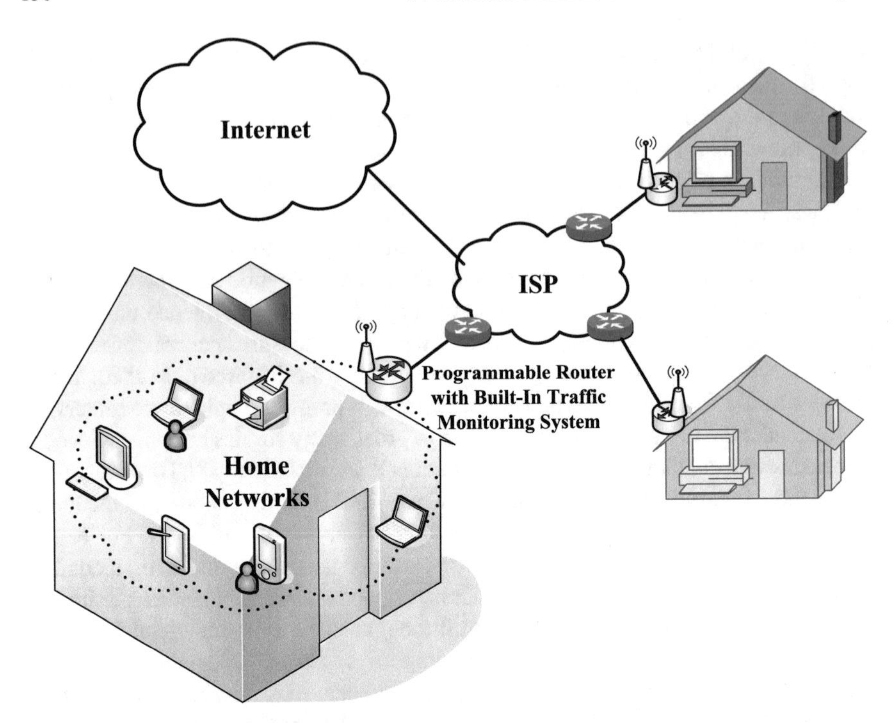

Fig. 8.4 Traffic monitoring platform for home networks

large number of home networks to demonstrate its benefits in managing and securing home networks.

Towards this end, we first present a traffic monitoring platform that collects network flow streams via traffic profiling servers and programmable home routers that connect home networks and the Internet via home gateways such as DSL or cable modems. Using traffic data collected from real home networks, we will analyze traffic patterns of connected devices in home networks, and characterize the volume, behavior, and temporal features of home network traffic.

Managing and securing the increasingly complicated home networks has remained a significant challenge for many home users who often have little technical expertise in network management [40, 50]. Many open-source tools or commercial products are available today to detect Internet malwares such as virus and worms or to filter known attacks through firewalls and intrusion detection systems for home networks. Unfortunately, there exist few simple and intuitive tools that could offer insights on traffic behaviors of home network devices.

To understand what is happening in home networks, we develop a real-time behavior monitoring platform to collect and analyze network traffic for Internet-capable devices in the home. As illustrated in Fig. 8.4, the monitoring platform captures network traffic via programmable home routers, which connect home networks with the Internet through home gateways such as cable or DSL modems. Using a Linux distribution for embedded devices, OpenWrt [51], we configure a programmable home router and export network flows traversing through all the interfaces of the router to a traffic profiling server running in the same home network. The continuous network flows, aggregated from IP packets, contain a number of important features for our traffic analysis including the start and end time stamps, source IP address (`srcIP`), destination IP address (`dstIP`), source port number (`srcPort`), destination port number (`dstPort`), and protocol, packets, and bytes. Many host-based monitoring systems are also able to collect these traffic flows on individual devices, e.g., Windows and Linux machines, however such host-based approaches are very difficult to deploy across all the possible devices due to the high heterogeneity of Internet-capable devices in the home.

Compared with incoming and outgoing traffic of home networks, the overhead of transferring flow data from programmable home routers to traffic profiling servers is not significant. Figure 8.5 shows the overhead of collecting traffic data from programmable home routers (top figure), the bandwidth usages of outgoing traffic (middle figure) and incoming traffic (bottom figure) of one home network that deploys the platform. As shown in the top figure, the network flow data exported by home routers consumes less than 4Kbps bandwidth, which is much smaller than outgoing and incoming traffic illustrated in the middle and bottom graphs. In general, the network bandwidth usage of incoming traffic towards home networks is larger than that of outgoing traffic, as most of Internet activities in these home networks are Web browsing, email communications, and video streaming.

The availability of the traffic monitoring platform makes it possible for us to analyze data traffic exchanged between home devices and Internet end hosts, as well as data traffic exchanged among home network devices. Making sense of these traffic could not only assist home users in understanding *what is happening in home networks*, but also help detect anomalous traffic towards home networks or originating from compromised home devices. In the next section, we will use traffic data collected from real home networks that deploy the traffic monitoring platform to characterize network traffic of Internet-capable home devices from a variety of traffic information including *volume features* measured by the numbers of flows, packets, and bytes, *social features* through analyzing IP addresses and application ports, and *temporal dynamics* of these traffic. Each of these traffic features captures the behavior of home devices from a unique perspective. Combined together, they provide a broad picture of home network traffic, and more importantly, reveal interesting traffic activities in home networks.

8.2.3 Characterizing Home Network Traffic

In this section, we first describe datasets used in this study and present the general characteristics of home network traffic. Subsequently, we explore principal component analysis to analyze temporal correlations among application ports for uncovering clusters of application ports sharing significant temporal patterns in network traffic.

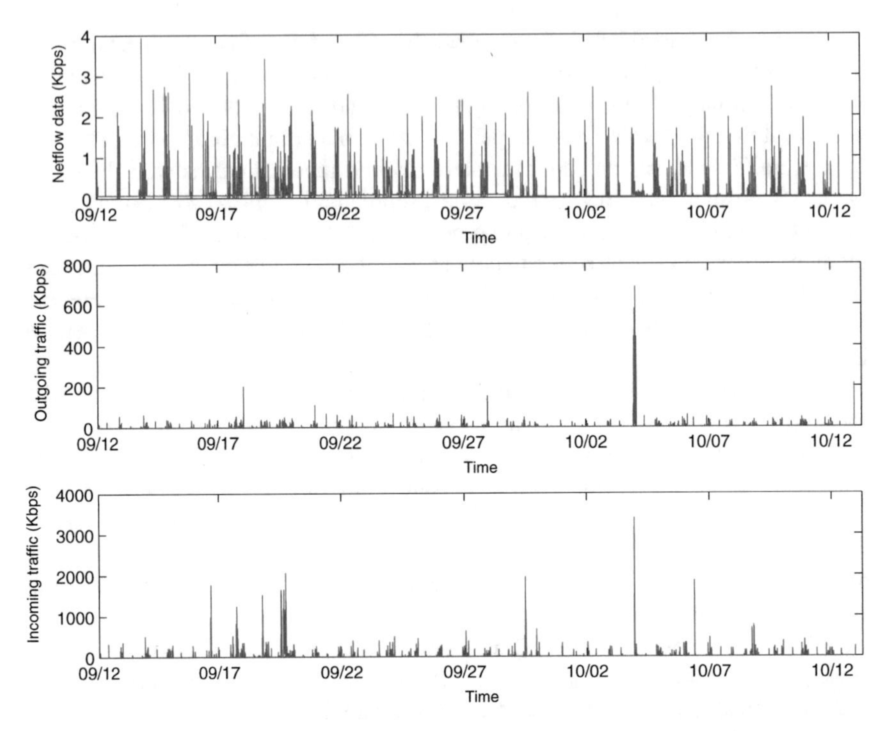

Fig. 8.5 Bandwidth usage of data collection (top figure), outgoing traffic (middle figure) and incoming traffic (bottom figure) of home networks

8.2.3.1 Datasets

The traffic data used in this study is collected from five home networks (A, B, C, D, E) that deploy our traffic monitoring platform during one-month time span from `09/12/2011` to `10/12/2011`. The numbers of total devices in home networks A and B are 6 and 3, respectively. Figure 8.6 shows the number of *online* devices in home network A over time. As illustrated in Fig. 8.6, the number of *online* devices in home network A observed during 5-min time bins varies from 0 to 6, reflecting Internet usage patterns of these devices during this one-month time period. Note that the number of home devices remaining above 1 between 09/12 and 09/28 is due to a probing program continuously running on one home device to measure end-to-end performance to a number of distributed servers. These devices collectively have communicated with over 4, 800 unique end hosts on the Internet from 529 different autonomous systems (ASes) during this period. Similarly, the devices in home network B collectively communicate with over 4, 400 end hosts from 726 ASes.

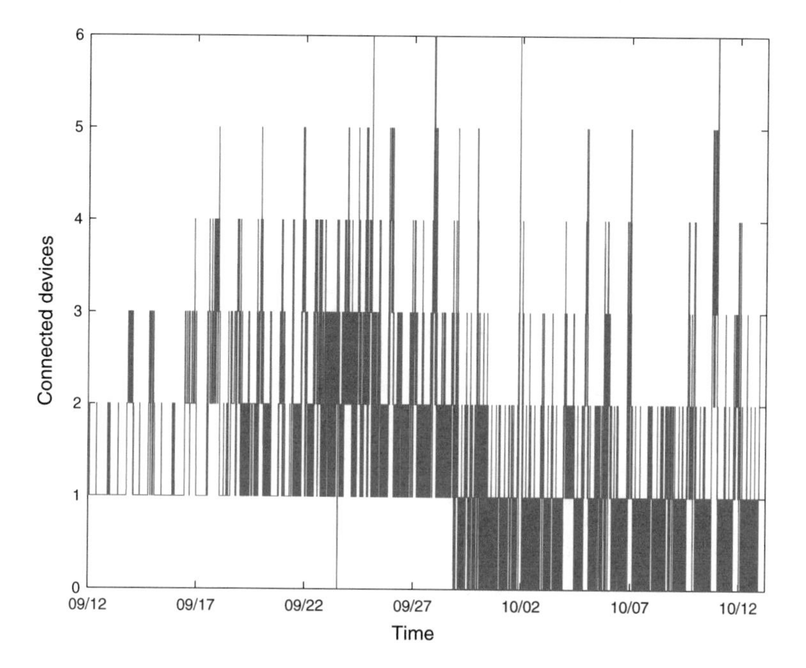

Fig. 8.6 The number of *online* devices in the home network A over time

8.2.3.2 Traffic Characteristics

We study the traffic characteristics of home networks by firstly examining IP addresses and application ports over time, since they reflect *whom do home devices communicate with* and *what applications do home devices use*. Figure 8.7a–c illustrate the numbers of unique destination IP addresses, unique source ports and unique destination ports for the outgoing traffic during 5-min time bins over time, respectively.

Our first interesting observation lies in the large number of unique destination IP addresses during 5-min time bins, as shown in Fig. 8.7a. Closer examination revealed that a single visit to a major content-rich Web portal could trigger tens of TCP connections to different Web servers, and the large number of destination IP addresses actually correspond to legitimate Web servers visited by home users. For example, our empirical experiment of visiting the front page of www.cnn.com with a Firefox browser finds that loading the entire page requires the browser to talk with 18 different IP addresses from a variety of Internet service and content providers including `Facebook` (social network site), `Google` (search engine), `Limelight Networks` (content deliver network), `Rackspace Hosting` (cloud service provider), `Valueclick` (online advertising), and `cnn` itself.

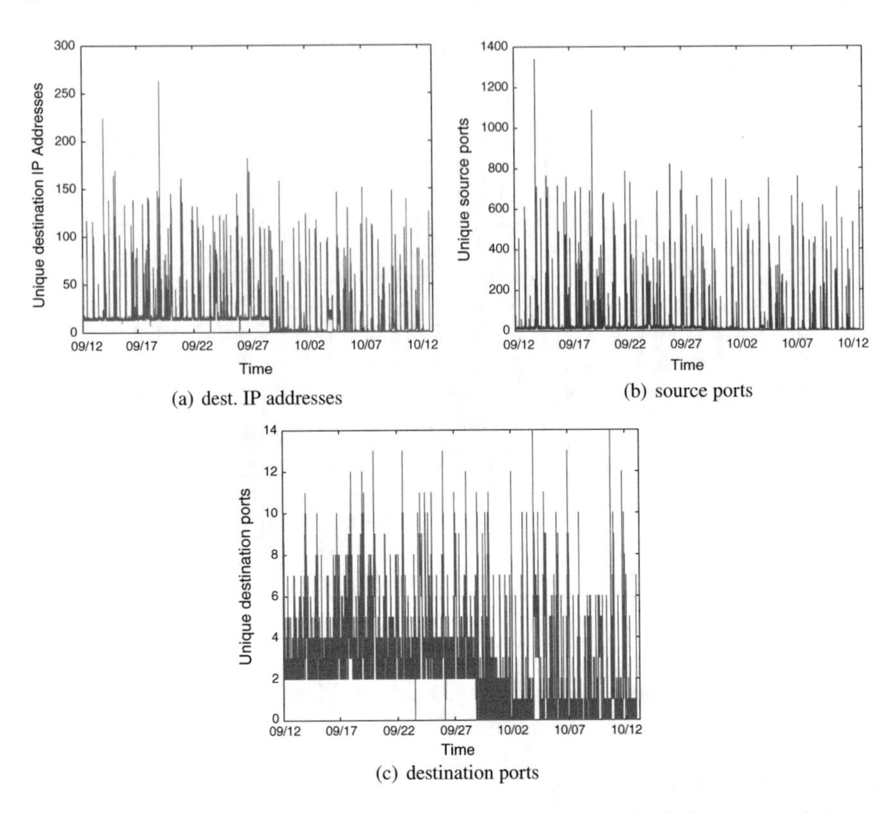

(a) dest. IP addresses

(b) source ports

(c) destination ports

Fig. 8.7 The number of unique IP addresses and ports in outgoing traffic for home network *A* over time

The second observation from Fig. 8.7 is that the number of unique destination ports for outgoing traffic in home networks is far less than that of unique source ports. The small number of destination ports in outgoing traffic provides a simple and natural classification on home network traffic, thus we follow a port-driven approach for further traffic analysis. Specifically, we separate outgoing traffic flows into distinct groups based on their destination ports in order to gain an in-depth understanding on network traffic of each individual destination port. Similarly, we group incoming traffic flows into distinct groups based on their source destination ports.

Figure 8.8a, b illustrate the temporal frequency of all destination ports for five home networks during one-month time period, respectively. It is interesting to find three types of temporal patterns among these ports. The first type of destination ports is consistently observed during all days. For example, port 80/TCP is observed in all days during the one-month period in both networks. The second type of ports is observed during several days, while the last type includes ports that are only observed in one or two days suggesting these infrequent ports might be associated with unusual or anomalous traffic. Similar observations hold for the source ports in the incoming traffic towards home networks. More interestingly, Fig. 8.8a, b also reveal temporal

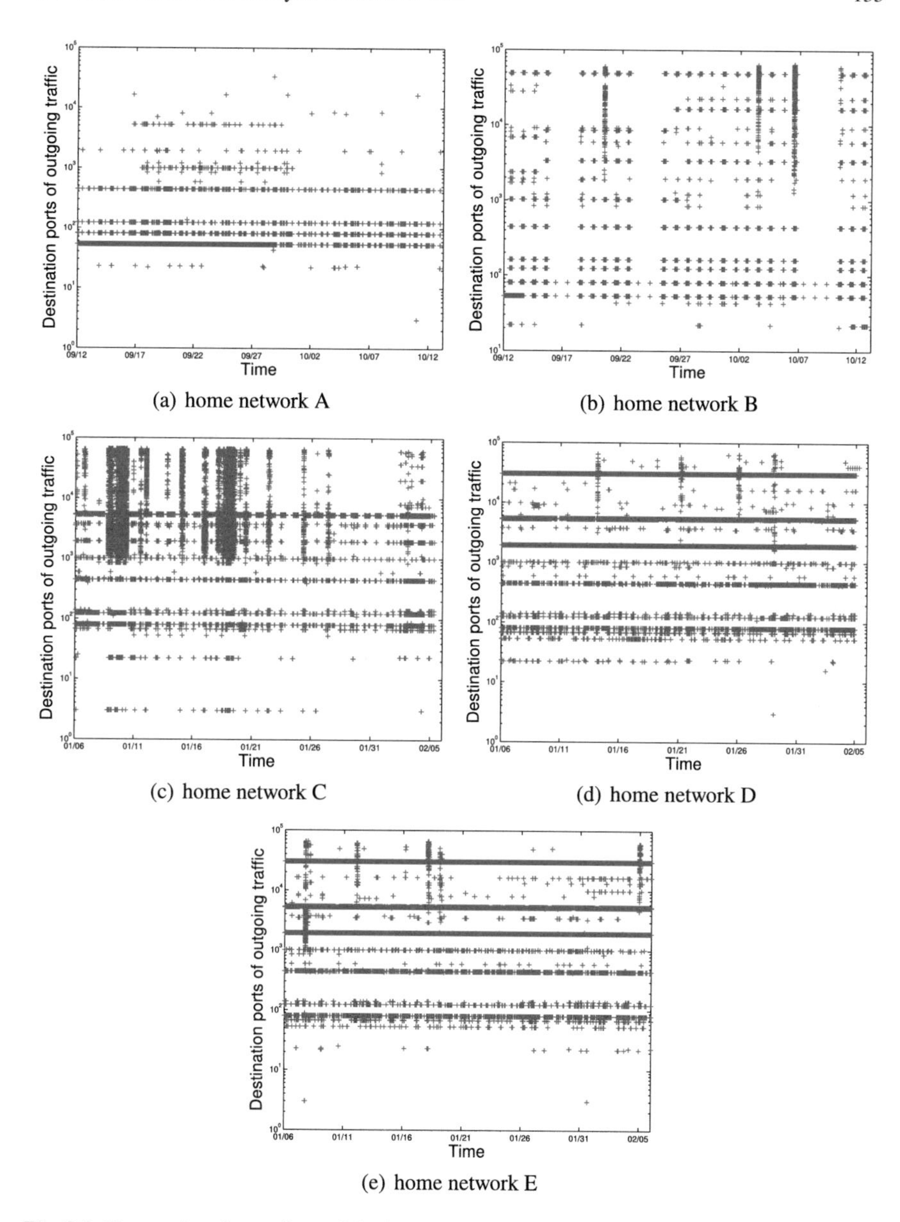

(a) home network A

(b) home network B

(c) home network C

(d) home network D

(e) home network E

Fig. 8.8 Time series observations of destination ports in outgoing traffic for five home networks

correlations among groups of applications ports that consistently show up around approximately the same times. This observation motivates us to explore correlation analysis techniques to understand the reasons behind such temporal correlations.

8.2.3.3 Temporal Correlation Analysis of Application Ports

To explore temporal correlation among application ports in home networks, we propose to use principal component analysis (PCA) to analyze traffic patterns of network applications. PCA is a widely used technique in network traffic analysis [52–55] due to its ability of analyzing multivariate data and locating interrelated variables [56].

Let p and t denote the total number of ports observed in the data and the total number of time bins. Our initial step is to construct a $p \times t$ matrix X, where $x_{i,j}$ denotes the total number of network flows for the destination port i ($i = 1, 2, \ldots, p$) in the outgoing traffic (or the source port i in the incoming traffic) during the j-th ($j = 1, 2, \ldots, t$) time period. The vector ${x_i}^T$ reflects a time series of observations for the application port i. Next we obtain the covariance matrix S, p non-decreasing ordered eigenvalues, $\lambda_1, \lambda_2, \ldots, \lambda_p$, and the corresponding eigenvectors $\alpha_1, \alpha_2, \ldots, \alpha_p$, where s_{ab} is the covariance of two application ports a and b, and $S\alpha_i = \lambda_i \alpha_i$, for $1 \leq i \leq p$.

The p principal components of the matrix X can be derived by projecting the matrix onto the p eigenvectors, i.e., $PC_i = \alpha_i^T X, i = 1, 2, \ldots, p$. As $var(PC_i) = var(\alpha_i^T X) = \alpha_i^T X \cdot X^T \alpha_i = \alpha_i^T S\alpha_i = \lambda \alpha_i^T \alpha_i = \lambda_i$, the variance captured by the i-th principal component is essentially the i-th eigenvalue λ_i.

PCA transforms the space of the p observed variables in the original matrix X into a new space of p principal components $\{PC_i\}, i = 1, 2, \ldots, p$. Figure 8.9 shows the distribution of the eigenvalues using the matrix constructed with the one-month traffic data from home network A. As shown in Fig. 8.9, a few largest eigenvalues account for the majority of the variance in the original matrix, suggesting that the corresponding top principal components capture most variances.

Thus, the final step of the PCA process is to project the original dataset onto a subspace with a smaller dimensionality to get approximate representations while retaining the majority of the variance in the original dataset. Specifically, we require that the largest m eigenvalues that are larger than a fixed threshold such that each selected principal component captures a non-trivial variance in the original datasets. In the experiment, we use 5% of the total variances as the threshold for determining the value of m.

The principal component PC_i can also be represented as $PC_i = \alpha_i^T X = [\alpha_{i1} x_1 + \cdots + \alpha_{ip} x_p]^T = [\sum_{j=1}^{p} \alpha_{ij} x_j]^T$, where α_{ij}, $j = 1, \ldots, p$, is the coefficient of x_j for PC_i. The coefficient value α_{ij} reflects the contribution or influence of the application port j to the variance obtained by the i-th component. Such relationship between principal components and observed variables leads to the discovery of a cluster of application ports that contribute similar influence towards the same principal components because of the inherent temporal correlations among these ports. As a result, we group the application ports that contribute similar high influence towards the variance of each of the top principal components into a distinct `srcPort` cluster for incoming traffic (or a `dstPort` cluster for outgoing traffic). In other words, PCA discovers the clusters of application ports that exhibit significant correlations in the temporal traffic patterns.

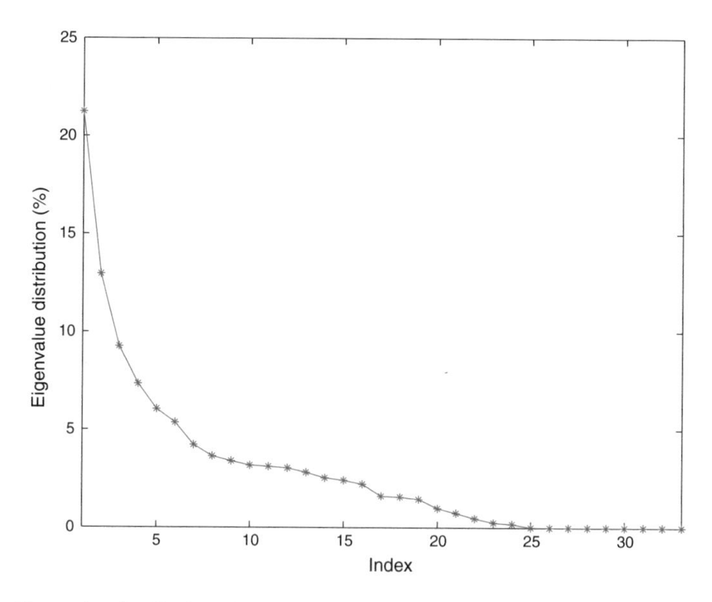

Fig. 8.9 Eigenvalue distribution of the matrix constructed with the one-month traffic data for home network *A*

Table 8.1 lists the membership of the 6 `dstPort` clusters discovered via the principal component analysis using one-month traffic data collected in home network *A*. $Cluster_1$ includes port 43/TCP and consecutive ports 33435-33440/UDP. The in-depth analysis shows that the flows associated with 43/TCP are legitimate *whois* traffic towards *Team Cymru IP to AS mapping service*, while all traffic associated with ports 33435-33440/UDP were sent towards an unknown server and failed to get a response from the server. The legitimate traffic on port 43/TCP and suspicious traffic on 33435-33440/UDP were observed during the same time window, which explain these seven ports to be grouped as a single `dstPort` cluster. Although $Cluster_1$ includes a service port 43/TCP, the majority of ports, 33435-33440/UDP, does reflect anomalous traffic activity from one home network device. $Cluster_2$ includes four canonical ports (i.e., DNS, HTTP, HTTPS, and NTP), which are used by home network devices on a daily basis and thus naturally form a `dstPort` cluster.

$Cluster_3$ includes three consecutive ports 16384-16386/UDP, which was sent by the FaceTime video calling application on an iPhone device. This cluster indicates that many user-installed applications or vendor-installed applications could use non-traditional ports for data communications with end hosts on the Internet. Such practices make it more challenging to differentiate anomalous or legitimate traffic on unusual ports. $Cluster_4$ includes three ports, i.e., 843/TCP, 1200-1201/TCP. Closer examinations reveal that a Windows laptop communicated with seven different instances in Amazon EC2 Cloud on these three ports *simultaneously* during 9 different days over the first two weeks. As home users are not aware of any application involving these ports and servers, these traffic is likely sent by a malware

Table 8.1 `dstPort` clusters discovered via PCA on temporal correlation

`dstPort` Cluster	Port Number	Application	User-aware
1	43/TCP	whois	Yes
	33435/UDP	unknown	No
	33436/UDP	unknown	No
	33437/UDP	unknown	No
	33438/UDP	unknown	No
	33439/UDP	unknown	No
	33440/UDP	unknown	No
2	53/UDP	DNS	Yes
	80/TCP	Web/HTTP	Yes
	123/UDP	NTP	Yes
	443/TCP	Web/HTTPS	Yes
3	16384/UDP	FaceTime	Yes
	16384/UDP	FaceTime	Yes
	16386/UDP	FaceTime	Yes
4	843/TCP	unknown	No
	1200/TCP	unknown	No
	1201/TCP	unknown	No
5	1863/TCP	MSN	Yes
	7001/UDP	MSN	Yes
6	993/TCP	IMAP over SSL	Yes
	5223/TCP	AppPush Notification Service	Yes

on the compromised laptop. $Cluster_5$ includes two ports 1863/TCP and 7001/UDP used by Windows MSN messenger, while $Cluster_6$ includes two ports 993/TCP and 5223/TCP, which are used by GMail and Apple Push Notification service running on the iPhone device that connects to the home network over Wi-Fi.

These experiment results with real home network traffic confirm that there indeed exist a variety of `dstPort` clusters that group applications ports with strong temporal correlations. Some of these clusters, e.g., $Cluster_1$ and $Cluster_4$ in Table 8.1, even lead to surprising findings on suspicious network traffic originating from home network devices that might be compromised by Internet malwares. Therefore, characterizing network traffic for Internet-capable devices in the home could not only provide valuable insight on behavior patterns of these connected devices, but also help improve the security and management of home networks.

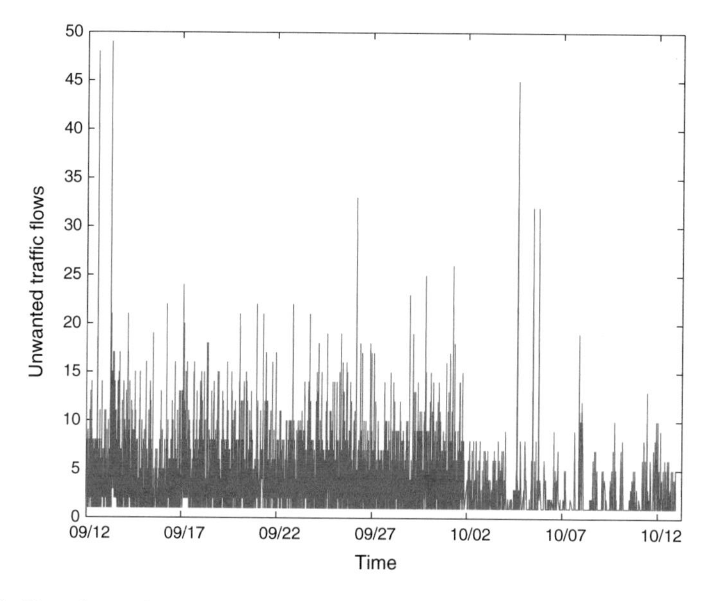

Fig. 8.10 The volume of unwanted traffic flows towards home networks over time

8.2.4 Unwanted Traffic Towards Home Networks

Unwanted or unproductive traffic is commonly observed on the Internet [23, 57]. Incoming network traffic collected from programmable home routers provides us an opportunity to characterize the intensity, sources, and port diversity of unwanted traffic towards home networks. In this study we consider an incoming traffic flow originating from the Internet as *unwanted* if none of Internet-capable devices in home devices responds to the flow.

Figure 8.10 illustrates the incessant activities of unwanted traffic towards home network A measured by the number of unwanted flows over time. Due to Internet background radiations [23, 57] formed by flooding backscatters, port scanning, and Internet worms, we consistently observe unwanted traffic from the Internet towards home networks. To gain a better understanding on where these traffic originates from, we analyze their source IP addresses and the corresponding ASes. Figure 8.11a shows the distribution of the source IP addresses of these unwanted traffic and their corresponding ASes in a log-log scale. As shown in Fig. 8.11a, there exist a few aggressive source IP addresses that send a fairly large number of unwanted traffic flows towards the same home network, while the majority of source IP addresses send only a few flows. The distribution for the source ASes exhibits similar heavy-tail patterns.

In addition to studying the sources of these unwanted traffic, we also characterize the diversity of source and destination ports in these traffic. Figure 8.11b shows the distribution of destination ports observed in unwanted traffic flows,

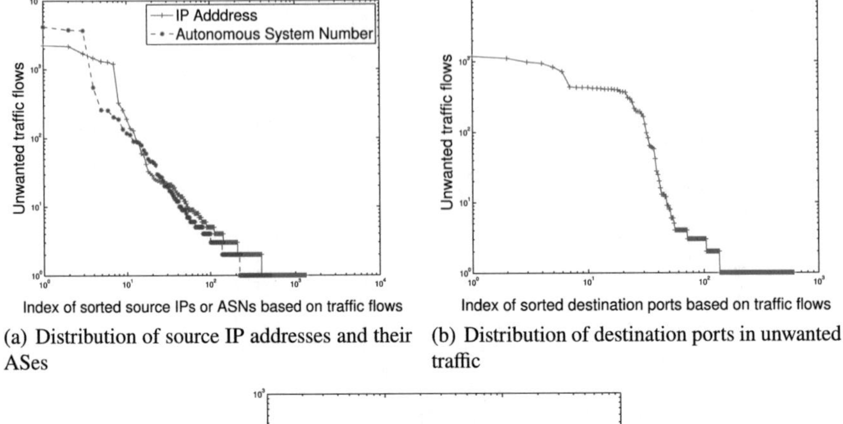

(a) Distribution of source IP addresses and their ASes

(b) Distribution of destination ports in unwanted traffic

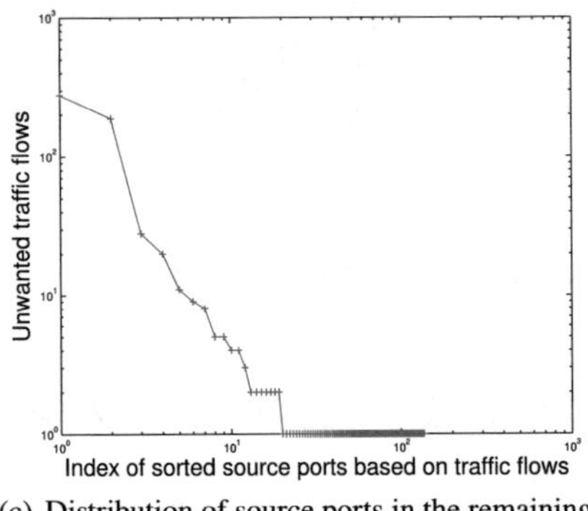

(c) Distribution of source ports in the remaining unwanted traffic

Fig. 8.11 Distribution of traffic features in unwanted traffic

and reveals that a small number of destination ports are observed with hundreds, or even thousands, of unwanted traffic flows. Most of these ports are associated with well-known vulnerability, such as port 1080/TCP (Mydoom), port 8000/TCP (a bot malware W32.Spybot.OGX) and port 9000/TCP (a mass-mailing worm W32.Mytob.GK@mm). The dominant nature on destination ports suggests that simple ACL (access control list) rules deployed at access routers of Internet service providers could become very effective for reducing a significant portion of unwanted traffic towards home networks. For traffic flows with unpopular destination ports, e.g., the number of flows less than 5, we aggregate these remaining flows on source ports for exploring the possible concentrations on source ports. As shown in Fig. 8.11c, many of these remaining flows actually are associated with a few popular source ports caused by flooding backscatters, including port 20/TCP (SSH), port 80/TCP (HTTP), port 443/TCP (HTTPS), port 3389/TCP (trojan Backdoor.Win32.Agent.cdm), port 6667/TCP (IRC) and port 12200/TCP (a recent malware targeting home routers).

The discovery of incoming unwanted traffic towards home networks reflects the challenges of stopping continuous Internet background radiations due to backscatters, port scanning, Internet malwares and other anomalous activities. The firewalls deployed in home routers are typically very efficient in filtering the majority of such unwanted traffic. However, these malicious traffic towards home routers could become successful if the routers were not properly configured or secured. In other words, protecting home routers and connected devices in the home becomes increasingly critical, as the compromised home devices and routers controlled by attackers would in turn become stepping stones to attack other vulnerable devices on the Internet. Uncovering root causes of how the home network devices exactly get compromised requires packet-level data, which is not collected by our traffic monitoring platform due to storage overhead on the storage-constraint home routers. However, our traffic monitoring platform has demonstrated the capability of detecting traffic patterns of malicious traffic that compromises the vulnerable devices in home networks. Extracting effective firewall rules from these traffic patterns and deploying these rules on home routers could filter such traffic, thus preventing home network devices from future attacks.

In summary, our experiment results confirm the wide spread of Internet malwares towards connected devices in home networks, and our analysis on unwanted traffic in home networks provides valuable input for botnets detection and spam filtering. As many Internet-capable devices in the home are likely compromised due to widespread Internet malwares, it is very important to develop effective and user-friendly tools to assist home users in properly configuring firewalls on home routers and end hosts to filter unwanted traffic from the Internet. Moreover, the prevalence of unwanted traffic in home networks calls for real-time traffic monitoring systems from inside home networks to enhance the overall Internet security.

8.3 Network Behavior Analysis for Internet of Things

The last decade has witnessed research advances and wide deployment of Internet of things (IoT) in smart homes and connected industry. However, the recent spate of cyber attacks exploiting the vulnerabilities and insufficient security management of IoT devices have created serious challenges for securing IoT devices and applications [58–60]. However, the recent spate of cyber attacks towards IoT devices in smart homes or small offices have created substantial challenges for Internet users without network and security expertise to manage and secure heterogeneous and poorly protected IoT devices [61–65]. As a first step towards understanding and mitigating diverse security threats of IoT devices, we present a measurement framework to automatically collect network traffic of IoT devices in edge networks, and build multidimensional behavioral profiles of these devices which characterize who, when, what, and why on the behavioral patterns of IoT devices based on continuously collected traffic data. In other words, the framework sheds light on the IP-spatial, temporal, and cloud service patterns of IoT devices in edge networks, and generates these multidimensional behavioral fingerprints for IoT device classification, anomaly traffic detection, and network security monitoring for millions of vulnerable and resource-constrained IoT devices on the Internet.

8.3.1 Background

The recent rapid development and deployment of IoT devices in smart homes, cities, and industry 4.0 have attracted significant interests from the research community in understanding the applications, security, threats, vulnerability, and the ecosystems [66–72]. IoT behavioral profiling and fingerprinting have recently attracted wide attention from the system, networking, and security research communities. The fingerprinting techniques cover nearly all protocol layers of TCP/IP stacks such as applying wavelet transform on the sequence of packet inter-arrival time (IAT) of wireless access points for device profiling [73–75] or characterizing packet headers and IP payload for device fingerprinting [76, 77]. Most of the existing studies on IoT behavioral fingerprinting are centered on the protocols of the physical and link layers for the applications of device classifications [73–75, 78]. For example, [74] introduces a real-time system that passively scans and analyze the data communication over Wi-Fi, Bluetooth, and Zigbee for classifying IoT devices and detecting privacy threats, while [75] proposes to extract the unique features from the link and service layers of Bluetooth low energy (BLE) protocol stack for generating the IoT fingerprint for authenticating devices and defending against spoofing attacks. In addition, [78] proposes a wireless device identification platform for distinguishing legitimate and adversarial IoT devices based on radio frequency (RF) fingerprinting over different ranges of signal-to-noise ratio (SNR) levels.

A few recent studies have shifted traffic data collection and analysis to the network, transport, and application layers for device behavioral modeling and characterizations [76, 77]. For example [77] establishes IoT device fingerprints with 20 binary features of protocol fields extracted from packets headers collected from link, network, transport, and application layers to reflect the protocol engagement of IoT devices headers such as ARP, IP, ICMP, TCP, UDP, NTP, DNS, DHCP, HTTP, and HTPPS, and 3 numerical features including packet size, destination IP counter, source, and destination port numbers, while [76] characterizes the behavioral fingerprints of IoT devices with a subset of binary features identified in [77], and 3 payload-based features including the entropy of payload, TCP payload size, and TCP window size. Compliment to these studies, the research on network behavior analysis for IoT devices in [79, 80] focuses on the behavioral fingerprinting of IoT devices in edge networks based network flow records, rather than the raw IP data packets which raise on privacy concerns of IoT users and computational resources on edge routers, for detecting new devices and traffic anomalies.

As the rapid and wide adoption of IoT devices continue to accelerate in smart homes, cities, and industries, it becomes increasingly urgent to design and implement Internet traffic measurement platforms to effectively monitor, characterize, and profile communications patterns of IoT devices with remote end hosts on the Internet and local systems on the same edge networks. Towards this end, we present a systematic measurement framework for establishing multidimensional behavioral profiles of connected IoT devices based on a wide spectrum of traffic features from IP-spatial, temporal, and cloud dimensions. Based on real network traffic data collected from a variety of edge networks over a long time span, we have discovered a number of important findings on behavioral fingerprints of IoT devices. First, IoT devices typically communicate with cloud servers from a very small number of prefixes and ASNs, which belong to IoT manufactories, the cloud service providers, NTP service providers, public DNS service providers. Second, IoT devices often exhibit repeated and predictable traffic activities over time due to heart-beat signals between IoT devices and cloud servers. Lastly, unlike laptops, desktops, and smartphones, IoT devices often engage with a limited and common number of applications such as DNS, HTTPS, HTTP, and NTP. These behavioral fingerprints not only summarize communication patterns of IoT devices with end systems on the Internet, but also benefit a range of security applications for IoT devices such as anomaly traffic detection, IoT detection and classification, and network security monitoring. As the link layer fingerprint could compliment the existing behavioral fingerprinting framework based on traffic features collected from network, transport, and application layers, an important research task of extending this framework is to explore the traffic fingerprints at the link layer, i.e., studying wireless communications between IoT hubs and IoT sensors via Bluetooth, ZigBee, Z-Wave, and Wi-Fi.

8.3.2 IoT Traffic Measurement and Monitoring

The burgeoning and insecure IoT devices in millions of edge networks call for effective techniques to detect, recognize, characterize, and address security threats towards these devices and applications. As a first step of securing IoT devices in edge networks, this chapter presents a measurement framework to automatically collect, process, characterize, and *profile* communication patterns of IoT devices with a variety of traffic features from IP-spatial, temporal, and service dimensions. Specifically, we leverage intelligent and programmable edge routers with commodity hardware to continuously collect incoming and outgoing network flow traffic in real-time for connected IoT devices in distributed edge networks.

The availability of network traffic data makes it possible to develop multidimensional traffic profiles of IoT devices for gaining an in-depth understanding of communication patterns and traffic behaviors of IoT devices, and more importantly, detecting and mitigating suspicious activities and cyber attacks towards vulnerable IoT devices. The additional benefit of measuring and monitoring network traffic of IoT devices is to have the full visibility of data communications and network configurations of IoT devices, e.g., Chromecast, a streaming media player developed by Google, configuring Google DNS servers as default rather than using the local ISP's DNS servers [81]. Such bogus behaviors are very hard to discover if the measurement functions are not available on home routers for capturing and profiling traffic activities of IoT devices in edge networks.

In this study we build the behavioral profile of IoT devices from a wide spectrum of their traffic features based on three dimensions: *IP-spatial*, *temporal*, and *cloud*. The IP-spatial dimension is centered on the analysis of remote IP addresses of Internet end hosts such as domain name system (DNS) servers or network time protocol (NTP) servers which IoT devices have communicated with. In addition, aggregating these remote IP addresses into Border Gateway Protocol (BGP) network prefixes [82] and ASNs allows us to analyze IP-spatial correlations of Internet end hosts communicating with IoT devices. Our experimental results on IP-spatial behaviors of deployed IoT devices in the wild have discovered that most IoT devices engage with cloud servers from a small set of network prefixes and ASNs due to their single-purpose applications and specific functions. For example, our experiment study discovers Philips Hue smart light bulbs mostly communicate with cloud servers, which are owned by Philips and deployed on Google cloud platforms, via Philips Hue smart hub for sending *on* or *off* commands.

Our proposed measurement framework characterizes behavioral profiles of IoT devices from the temporal dimension through identifying three distinct temporal traffic patterns from connected objects in edge networks, and classify IoT devices into always-on and on-demand devices. For the analysis on the cloud dimension, our study shows that IoT devices typically only engage with a small and fixed set of common applications such as Hypertext Transfer Protocol (HTTP), DNS, and NTP due to their specific functionalities.

In light of the prevalent cybersecurity threats against IoT devices in edge networks, we explore the benefits of multidimensional behavioral profiles for a wide spectrum of applications including anomaly traffic detection, IoT device detection, and classification, and network security monitoring. Specifically, we introduce a simple yet effective pattern-based anomaly detection approach for encoding common network traffic patterns with short encoded length, and encoding infrequent and unusual patterns with longer encoded length. The experimental evaluation shows that the approach is able to uncover suspicious traffic activities with high precisions. Moreover, we leverage multidimensional profiles of IoT devices for recognizing and detecting new and unknown IoT devices based on the profiles of existing and known IoT devices. Finally we outline how the behavioral profiles could facilitate network security monitoring via effectively capturing behavioral dynamics or deviations caused by cyber attacks such as port scanning activities and repeated failed login attempts.

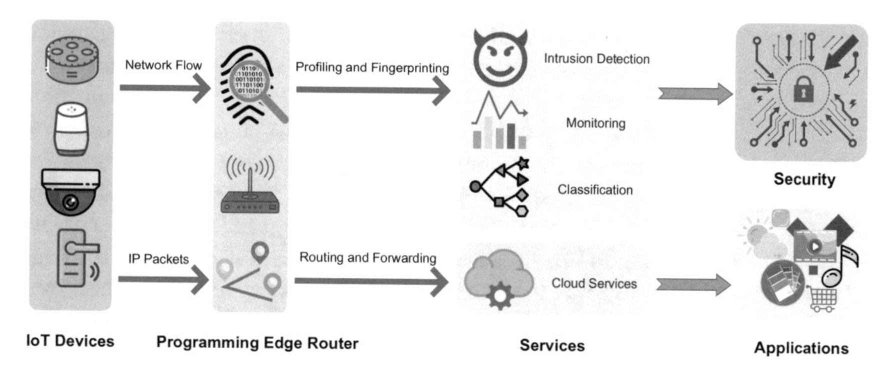

Fig. 8.12 An IoT traffic measurement framework via programmable routers at edge networks

8.3.3 An IoT Traffic Measurement Framework via Programmable Edge Routers

Recent advances on embedded systems, sensors, robotics, and machine learning have enabled the wide deployment of IoT devices in edge networks. The first step of protecting and securing millions of IoT devices is to measure, monitor, and understand their normal communication patterns and behavioral profiles. For example, what remote hosts on the Internet are talking with the smart speakers or thermostats at home networks, at what time, for what reasons? A recent security evaluation study [83] on IoT deployment has also pointed out that measurement is a crucial step for protecting the security of IoT devices and the privacy of end users.

Answering these questions is very critical to understand if and when connected IoT devices in edge networks are compromised by cyber attacks such as Mirai botnet [84]. The Mirai botnet has successfully infected over 60,000 IoT devices including IP cameras and consumer-grade routers in the first 20 h after being released to the Internet, and launched more than 15,000 cyber attacks towards game servers, telecoms, anti-DDoS providers, and other high-profile Web sites.

Towards profiling communication patterns of IoT devices, we leverage the computational resources on intelligent and programmable edge routers to develop a prototype measurement framework, which is able to capture network traffic flows of IoT devices for real-time traffic monitoring and behavioral profiling. As shown in Fig. 8.12, the programmable edge router continuously captures, stores, and analyzes the incoming, outgoing, and internal network traffic flow records of all IoT devices in the edge network. For each flow record, our measurement framework collects the well-known 5-tuples of a network conversation or session, i.e., source IP address (`srcIP`), source port number (`srcPort`), destination IP address (`dstIP`), destination port number (`dstPort`), and protocol, as well as the start and end time stamps, byte count, and packet count.

Our measurement framework does not collect raw IP packets from IoT devices since most data packets originating from or destined to IoT devices are encrypted, and the storage of raw data packets of IoT devices such as smart TVs or IP cameras could bring undesired system challenges for resource-constrained edge routers. On the other hand, network flow records are widely used for Internet traffic classification, network measurement and analysis [85, 86] thanks to their diverse and informative traffic features and marginal computational and storage resource overheads.

In this study, we have collected network flow records of IoT devices from 22 home networks and small offices in the United States, Hong Kong, and China. The number of end systems including IoT devices and non-IoT devices connecting to each edge network ranges from 1 to 25. In total, these edge networks collectively connect over 50 IoT devices including Amazon Echo, Google Home, Philips Hue smart light bulbs, Samsung smart plug and motion sensor, YI home camera, August smart lock, LG smart TV, and a number of other IoT devices. To demonstrate the practical feasibility of the IoT traffic measurement framework, we deploy and evaluate the system with different brands of programmable routers including Linksys, Netgear, Buffalo, and CanaKit Raspberry Pi.

8.3.4 Multidimensional Behavioral Profiling of IoT Devices

In this section we present a multidimensional behavioral profiling approach for fingerprinting the behaviors of IoT devices from a wide spectrum of traffic features based on network flow records collected from edge networks. First, we study the *IP-spatial* behavior of IoT devices via characterizing remote IP addresses engaging with IoT devices and aggregating these IP addresses into BGP networks prefixes and ASNs for correlation analysis. Subsequently, we study the *temporal* traffic patterns of IoT devices over our longitudinal measurement study, and profile the cloud behaviors of IoT devices via analyzing how they interact with cloud servers.

8.3.4.1 IP-Spatial Behavior of IoT Devices

We characterize the IP-spatial behaviors of IoT devices by analyzing the remote IP addresses which communicate with these devices. More importantly, we aggregate and correlate these remote addresses into BGP network prefixes and ASNs for gaining an in-depth understanding of "clustered" IP-spatial behaviors for IoT devices. For example, the IP address of the DNS server for Google home smart voice assistant, `8.8.8.8`, is from the BGP prefix `8.0.0.0/9` and ASN 15169 owned by Google based on the latest snapshot of the BGP routing table [87] and the official registry records from Internet assigned numbers authority (IANA).

Aggregating and correlating remote individual IP addresses to network prefixes and ASNs reveal an interesting observation. IoT devices typically engage with a very small subset of BGP network prefixes and ASNs, even though they communicate with a large number of remote severs, which are likely from the same server pool by the same service providers for efficient load balancing and content distributions.

Table 8.2 The clustered patterns of IP-spatial behavior of IoT devices in the same edge network during a 5-min time window

Device	IoT	`dstIPs`	prefixes	ASNs
Amazon Echo	Yes	3	3	1
Echo Dot	Yes	5	4	1
IP Camera	Yes	2	2	1
Philips Hue	Yes	1	1	1
Samsung smart plug	Yes	3	2	1
Smart TV	Yes	4	3	2
Smart Phone	No	37	24	13
Laptop	No	172	102	39

Table 8.2 summarizes the clustered patterns of IP-spatial behavior of 6 IoT devices and 2 non-IoT devices in one edge network during a 5-min time window. As shown in Table 8.2, each IoT device only engages with servers from one or two *unique* ASNs during the observation period, while the smartphone and laptop communicate with remote end hosts from 13 and 39 *unique* ASNs, respectively.

Figure 8.13 shows the convergence of unique remote IP addresses, their network prefixes, and ASNs for a variety of IoT and non-IoT devices in the same edge network over a 4-month time span. As shown in the longitudinal measurement study for the IP-spatial behavior, it is very interesting to observe that all IoT devices have engaged with a much smaller set of destination IP addresses, prefixes, and ASNs than smartphones and laptops.

8.3.4.2 Temporal Behavior of IoT Devices

For the temporal behavior of IoT devices, we first measure the number of distinct time slots in which IoT devices exhibit traffic activities during the longitudinal measurement study. In this study, we select 5 min as the time unit for analysis to balance the computation overhead and monitoring real-time traffic activities, thus the maximum of time slots an IoT device is observed is 288 in one day. Figure 8.14 shows the flow, packet, and byte counts of three different connected devices in edge networks over one-week time span. As shown in Fig. 8.14, the smart voice assistant, smart TV, and smartphone exhibit distinct traffic characteristics over time, and have very unique and diverse temporal patterns on flow, packet, and byte counts over time, which leads us to measure and quantify the *variability* on the number of time windows for IoT devices over the entire data collection period.

For each IoT device d in the edge network, let $t_{d,i}$ represent the number of time windows the device d is observed with network traffic on the i-th day. Considering connected devices are randomly added into the edge network, we use the average

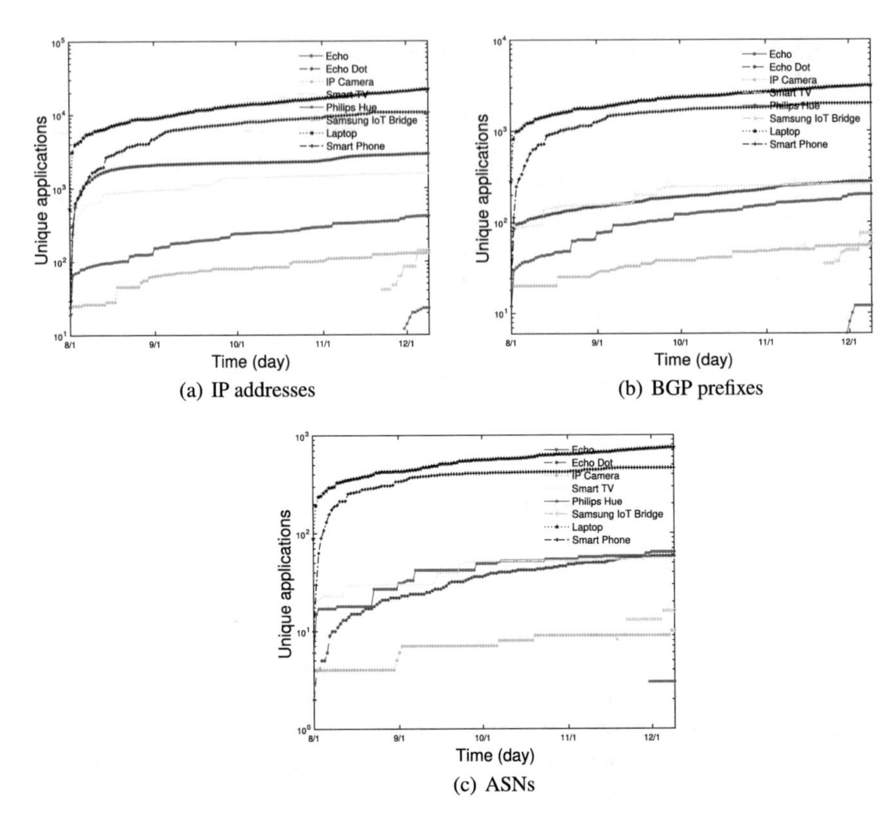

(a) IP addresses

(b) BGP prefixes

(c) ASNs

Fig. 8.13 The convergence of IP addresses, prefixes, and ASNs for IoT and non-IoT devices over the longitudinal measurement period

time window for each device μ_d rather than the total number of time windows during the entire measurement period. The average of time windows μ_d is derived as $\mu_d = \frac{\sum_{i=1}^{N} t_{d,i}}{N}$, where N is the number of the days since the device d is observed in the edge network and $1 \leq i \leq N$. Finally, the actual temporal variability on time windows, measured by coefficient of variance, is calculated as $CoV_d = \frac{\mu_d}{\sigma_d}$, where σ_d, the standard deviation, is calculated as $\sigma_d = \sqrt{\frac{1}{N}\sum_{i=1}^{N} t_{d,i} - \mu_d}$.

Figure 8.15 illustrates a scatter graph on the mean μ and coefficient of variance CoV of time slots observed with traffic activities for different IoT and non-IoT devices deployed in the same edge network. As shown in Fig. 8.15, four out of the six IoT devices exhibit traffic activities during the majority of time windows in each and every day, and their *variability* on the number of time windows is much smaller than that of non-IoT devices. One IoT device, i.e., an IP camera, is only active for a small number of time slots per day, but exhibits lows variability on the time window as well. The only IoT device showing a high variability on the number of time windows across different days is a smart TV, which is turned on and off in an

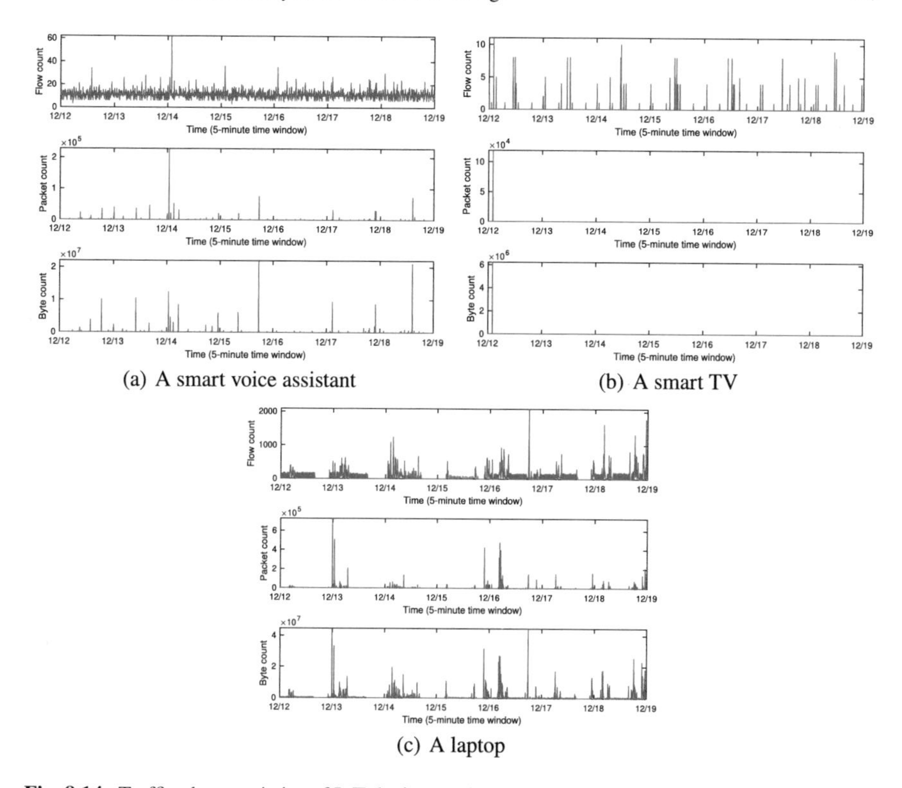

(a) A smart voice assistant

(b) A smart TV

(c) A laptop

Fig. 8.14 Traffic characteristics of IoT devices and non-IoT devices over 1-week time span

unpredictable fashion. Based on these observations on the temporal patterns of IoT and non-IoT devices, we can classify connected devices in edge networks in three categories: always-on IoT devices, on-demand devices, and non-IoT devices.

The self-similarity traffic patterns of IoT devices visualized on Fig. 8.14 also inspire us to analyze the autocorrelation on network traffic for all connected devices in edge networks. The autocorrelation metric quantifies the correlation of the same variable across different and lagged periods of times, thus the metric is also referred as to serial correlation and lagged correlation. The autocorrelation metric, $\rho_{d,k}$, for the IoT device d, between network traffic activity time series $X_{d,t}$ and a k-lagged copy of itself $X_{d,t+k}$ is captured by the autocorrelation function (ACF) as follows:

$$\rho_{d,k} = \frac{\sum_{t=k+1}^{n-k}(X_t - \mu)(X_{t+k} - \mu)}{\sigma^2}, \tag{8.1}$$

where μ and σ are the mean and standard deviation of network traffic activity time series X_d, respectively. An autocorrelation value of 0 suggests independent and random observations on the traffic time series of connected devices in edge networks,

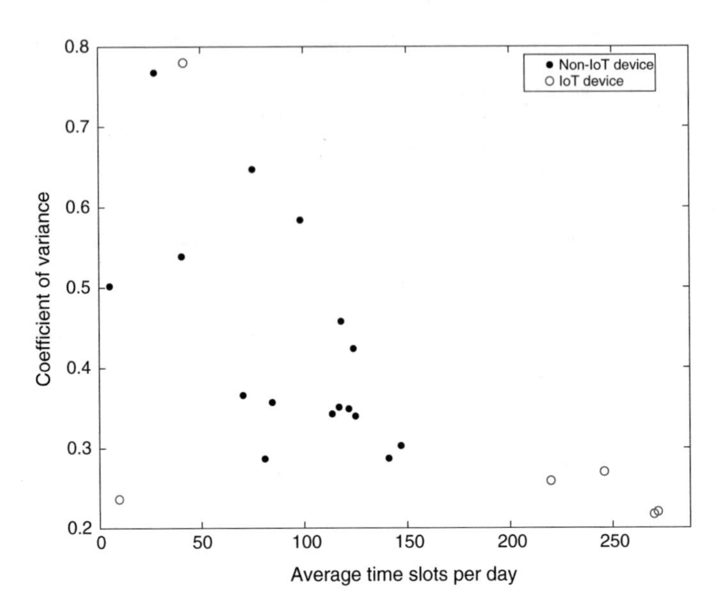

Fig. 8.15 The mean and coefficient of variance of time slots observed with traffic activities for IoT and non-IoT devices

while a significant autocorrelation reveals substantial correlations among adjacent observations or determines predictable seasonality in the time series [88, 89].

Figure 8.16 illustrates the autocorrelation plots, also referred to as correlograms, of network traffic time series for three selected IoT and non-IoT devices. As shown in Fig. 8.16, the network traffic time series of IoT devices in edge networks indeed exhibit various extents of self similarity patterns.

8.3.4.3 Cloud Behavior of IoT Devices

The objective of characterizing cloud behavior of IoT devices is to understand why IoT devices communicate with remote servers in the cloud. In particular, we profile cloud behaviors of IoT devices based on the *dominant* applications or services observed from `dstPort` and protocol of their outgoing network traffic flows. Table 8.3 illustrates all the observed 5 applications for the 6 IoT devices deployed in one edge network during a 24-h time window. These 5 applications are HTTP, Hyper Text Transfer Protocol Secure (HTTPS), DNS, NTP, and Spotify music streaming. As a comparison, one smartphone and one laptop in the same edge network engage with 11 and 15 distinct applications, respectively, during the same time period.

The limited and consistent set of common applications used by IoT devices confirms that IoT devices are typically designed for very specific functions and dedicated utilities. Figure 8.17 illustrates the convergence of cloud applications for IoT and non-IoT devices. As shown in Fig. 8.17, the number of applications for IoT devices

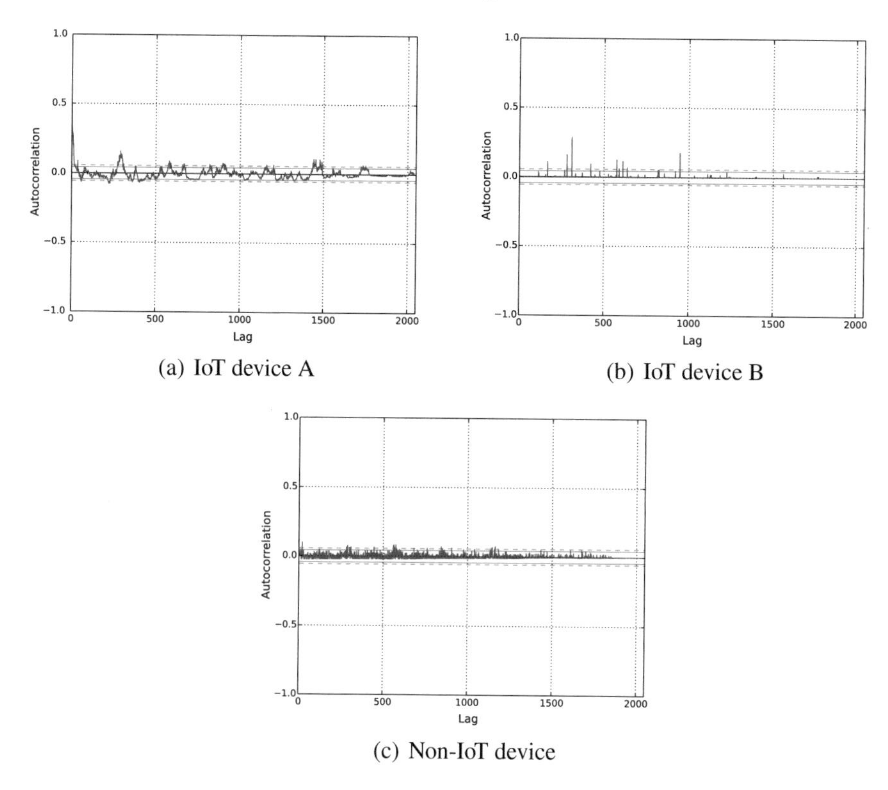

(a) IoT device A

(b) IoT device B

(c) Non-IoT device

Fig. 8.16 The autocorrelation plots of network traffic time series for selected IoT and non-IoT devices

Table 8.3 The dominant applications used by IoT devices in edge networks

Application	Service	Echo	Camera	Echo Dot	Philips Hue	Smart TV	IoT Hub
443/TCP	HTTPS	Y	Y	Y	Y	Y	Y
80/TCP	HTTP	Y		Y	Y	Y	Y
53/UDP	DNS	Y	Y	Y		Y	
123/UDP	NTP	Y	Y	Y	Y		
4070/TCP	Spotify	Y					

converges in a very rapid fashion. It is very interesting to note that all IoT devices use HTTPS for secure and encrypted Web services, which shows the security awareness and investment of IoT manufactories and application developers. On the other hand, the non-encrypted HTTP service is still observed for five IoT devices.

For each application, we continue to characterize the remote servers and their aggregated network prefixes or ASNs via analyzing the *fanouts*, i.e., unique numbers of destination IP address, BGP prefixes, and ASNs. In addition, we measure the

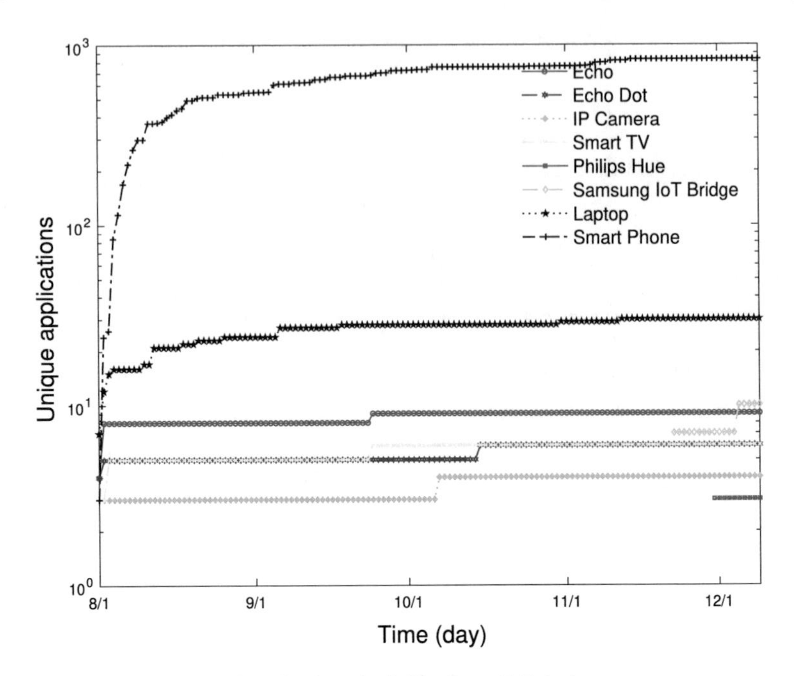

Fig. 8.17 The convergence of applications for IoT and non-IoT devices

distribution of network traffic across these remote servers, prefixes and ASNs via calculating the entropy and standardized entropy of these fanouts. For a given application a for an IoT device d, let N and m denote the number of network traffic flows and the *unique* number of the remote servers represented as $s_1, s_2, \ldots, s_m$. The probability of each remote server p_{s_i} is calculated as $p_{s_i} = \frac{f_{s_i}}{N}$, where f_{s_i} denotes the number of flows between d and s_i. Clearly $\sum f_{s_i} = N$. The entropy on the remote servers for the application a for the device d is then derived as $\mathcal{E}_{d,a} = -\sum_{i=1}^{m} p_{s_i} \log p_{s_i}$, while the normalized entropy is derived as $\mathcal{NE}_{d,a} = \frac{\mathcal{E}_{d,a}}{\log m}$.

The normalized entropy is in the range of [0, 1], revealing the degree of uncertainty, randomness, or variations on the remote servers which communicate with IoT devices in edge networks. Clearly, a $\mathcal{NE}_{d,a}$ value of 0 or near 0 indicates the uniformity on the remote servers, while a $\mathcal{NE}_{d,a}$ value of 1 or near 1 means the high randomness on the remote servers. The former scenario indicates the IoT device only communicates with one or a few servers on the application a, while the latter case reveals the device talking with a large number of random servers. Based on a similar process, we could calculate the entropies and normalized entropies for their aggregated network prefixes or ASNs of remote servers. Table 8.4 illustrates the entropy values of destination IP addresses, prefixes and ASNs IoT devices have sent HTTPS requests within a 24-h time window. As shown in Table 8.4, all IoT devices exhibit how uncertainty on network prefixes and ASNs for their HTTPS traffic, while the laptop and smartphones exhibit much higher variations on the remote prefixes and ASNs for HTTPS traffic.

Table 8.4 The entropy of destination IP addresses, prefixes, and ASNs IoT devices have sent HTTPS requests within a 24-h time window

Device	Flows	Fanout			Normalized Entropy		
		IP	Prefix	ASN	IP	Prefix	ASN
Echo	148	20	6	1	0.5529	0.3158	0.0000
Camera	32	12	9	2	0.6023	0.5422	0.1792
Echo Dot	228	40	10	2	0.6197	0.3365	0.0051
Philips Hue	96	4	2	1	0.2163	0.0221	0.0000
Smart TV	429	109	39	7	0.6574	0.2968	0.1733
IoT Hub	258	3	2	1	0.1969	0.1115	0.0000
Laptop	3831	832	340	90	0.6782	0.5191	0.3064
Smartphone	1497	353	131	21	0.6274	0.4964	0.3077

These observations could potentially provide critical insights for detecting traffic anomalies of IoT devices or classifying newly added IoT devices to the edge network. In summary, our multidimensional behavioral profiling of IoT devices have led to a number of discoveries. First, aggregating and correlating the remote IP addresses into BGP networks prefixes and ASNs reveal IoT devices typically engage with servers from a small number of networks and domains due to their specific and single-purchase functionalities. Second, the temporal traffic patterns could classify IoT devices into always-on devices such as smart voice assistants and on-demand devices such as smart TVs. Lastly, most IoT devices communicate with Internet servers for limited, fixed, and common applications such as HTTP, DNS, and NTP services. Profiling traffic behaviors of IoT devices not only uncover what, when and how IoT devices communicate with legitimate end hosts on the Internet, but also provide critical insights for detecting suspicious activities of IoT devices due to security threats and cyber attacks. Thus, the next section leverages IoT behavioral fingerprints for a wide variety of applications such as IoT device detection and classification, anomaly traffic detection, and cybersecurity monitoring.

8.3.5 *Exploring the Applications of Multidimensional Behavioral Profiling*

In this section, we demonstrate the benefits of multidimensional behavioral profiles of IoT devices for a variety of applications including anomaly traffic detection, IoT device detection and classification, and network security monitoring.

8.3.5.1 Anomaly Traffic Detection for IoT Devices

Security and privacy are two key challenges faced by today's wide deployment of IoT devices in edge networks due to inadequate built-in security features, flawed authorization and authentication processes, weak password management, and other vulnerabilities. As cyber attacks exploring millions of weakly protected IoT devices often leave substantial traffic footprints in edge networks, we explore multidimensional behavioral profiles for detecting anomaly traffic and security threats.

In this study, we adopt an anomaly detection method based on minimum description length (MDL) principle due to its data-driven approach and parameter-free feature [90–92]. The intuition and novelty of the MDL principle lie in its pattern-based compression and encoding technique which exploit coding tables to capture the underlying data distributions. In other words, the technique encodes a frequent and common pattern with a short encoded length, and encodes a less frequent and unusual pattern with a long encoded length reflecting anomalies and irregularities in the original data [90].

The MDL principle essentially is a model selection framework for performing lossless compressions and encoding on data with categorical features and attributes. The main process is to search and identify the best model m which minimizes the overall encoding size for the entire data, i.e.,

$$\arg\min_{m \in \mathcal{M}} L(m) + L(d \mid m), \tag{8.2}$$

where $\mathcal{M}$, $L(m)$, $L(d \mid m)$ are the model set, the bit length describing the specific model m, and the bit length of describing the data d with the model m, respectively.

In the context of network flow traffic of IoT devices in edge networks, we consider all network flow data collected during a given time period as the dataset D consisting of n flow records, each of which has w categorical features, i.e., $\mathcal{F} = \{f_1, \ldots, f_w\}$. To encode the data with a code table, CT, we first extract all the patterns $\mathcal{P}$ in the data, and represent each pattern with a code c in the encoding set C. For a given pattern $p \in \mathcal{P}$ encoded as $c(p)$, we define its frequency, i.e., $freq(p)$ as the number of flow records in D containing p in their encoding. Thus based on the entropy theory, the optimal coding for the pattern p becomes

$$L(c(p) \mid CT) = -\log\left(\frac{freq(p)}{\sum_{q \in CT} freq(q)}\right).$$

In addition, the overall number of bits required to encode the entire dataset D is derived as

$$\begin{aligned} L(D \mid CT) &= \sum_{r \in D} L(r \mid CT) \\ &= \sum_{r \in D} \sum_{p \in freq(r)} L(c(p) \mid CT). \end{aligned}$$

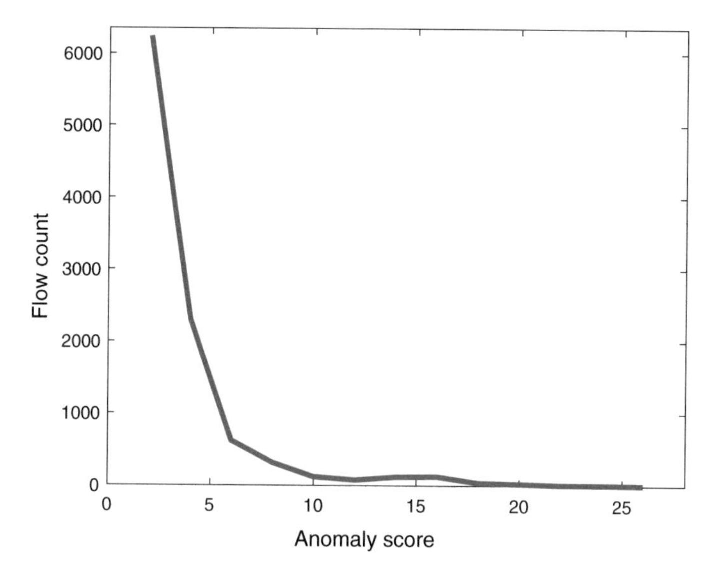

Fig. 8.18 The distribution of anomaly scores for all observed network traffic flows during a 24-h time window for a smart voice assistant

As shown in Eq. 8.2, the bit length of encoding the overall data is then calculated as

$$L(CT) = \sum_{p \in CT} L(c(p) \mid CT) + \sum_{v \in \mathcal{V}} -o_v \log(p_v),$$

where $\mathcal{V}$ is the set of all unique categorical attributes appearing in the patterns of the code table, o_v is the occurrence count of the category value $v \in \mathcal{V}$. p_i is calculated as $\frac{o_i}{\mathcal{L}}$ where $\mathcal{L}$ is the total length of all the patterns in the code table. Combining the entire feature set together, we can build multiple code tables for further reducing the overall encoding cost.

The simple yet effective pattern-based anomaly detection approach allows us to identify unusual or anomalous traffic flows from network traffic originating from or destined to IoT devices in edge networks. Our encoding process leverages the following multidimensional traffic features extracted from network flow records: flow duration, `srcIP`, `srcPort`, `dstIP`, `dstPort`, protocol, packet count, byte count, `dstIP`'s network prefix, and `dstIP`'s ASN. The MDL principle intends to encode unusual patterns with longer encoded lengths, thus we simply consider the encoding length $L(r \mid CT)$ for a network flow record r as the anomaly score.

Figure 8.18 illustrates the distribution of anomaly scores for all the observed network traffic flows originating from a Google Home smart voice assistant during a 24-h time window. Based on the widely used elbow principle, we determine the anomaly score of 9 as the threshold for traffic anomalies for IoT devices in edge networks. To evaluate the quality of the anomaly detection, we manually validate all 526 network flows with an anomaly score of 9 or above.

Table 8.5 An in-depth analysis of network traffic flows high anomaly scores

Protocols	Root cause analysis	Flows
HTTPS	long secure web sessions with cloud servers	489
ICMP	ping traffic	13
mDNS	multicast DNS query	3
DHCP	DHCP requests	9
DNS	Unusual number of Packets	2
8009/TCP	Optimized HTTP service running on the device.	2
5228/TCP	long TCP connections with Google Play services	8

Table 8.5 summarizes our in-depth analysis of all 526 network flows with high anomaly scores. As shown in Table 8.5, most of these network flows are long HTTPS connections between the smart voice assistant with Google cloud servers. In addition, a small number of network flows are related to ICMP, mDNS, and DHCP traffic. Thus the manual validation confirms the effectiveness of our proposed pattern-based anomaly detection for discovering unusual traffic activities from the multidimensional behavioral profiles of IoT devices.

8.3.5.2 IoT Device Detection and Classification

The multidimensional behavioral profiles of existing IoT devices in edge networks also provide unique and valuable features for detecting and classifying newly added devices to the network. Let i and j denote two IoT devices in the dataset. For each and every traffic feature in behavioral profiles over a given time window, we can quantify and measure the similarity and correlations of the feature between two devices i and j during the same time period. Assuming the feature b is the remote destination IP addresses (`dstIPs`) that communicate with IoT devices. Let $\mathcal{S}_{i,b}$ and $\mathcal{S}_{j,b}$ represent the unique sets of `dstIPs` observed for IoT devices i and j during the time window, respectively. The similarity on the `dstIP` feature, i.e., $s_{i,j,b}$, is calculated as

$$s_{i,j,b} = \frac{|\mathcal{S}_{i,b} \cap \mathcal{S}_{j,b}|}{|\mathcal{S}_{i,b} \cup \mathcal{S}_{i,b}|}. \tag{8.3}$$

Thus repeating the same process on the available features extracted from network flow data could lead to a *similarity vector* for any two IoT devices in the same or different edge networks. The similarity matrix on traffic features among all IoT devices enables us to identify and cluster devices with similar behavioral fingerprints, and more importantly detect new suspicious IoT devices in the same edge network.

Figure 8.19 illustrates the distributions of similarity scores on three IP-spatial features including `dstIP`, destination prefixes and ASNs between IoT devices in two different edge networks. Each point represents one pair of IoT devices from two

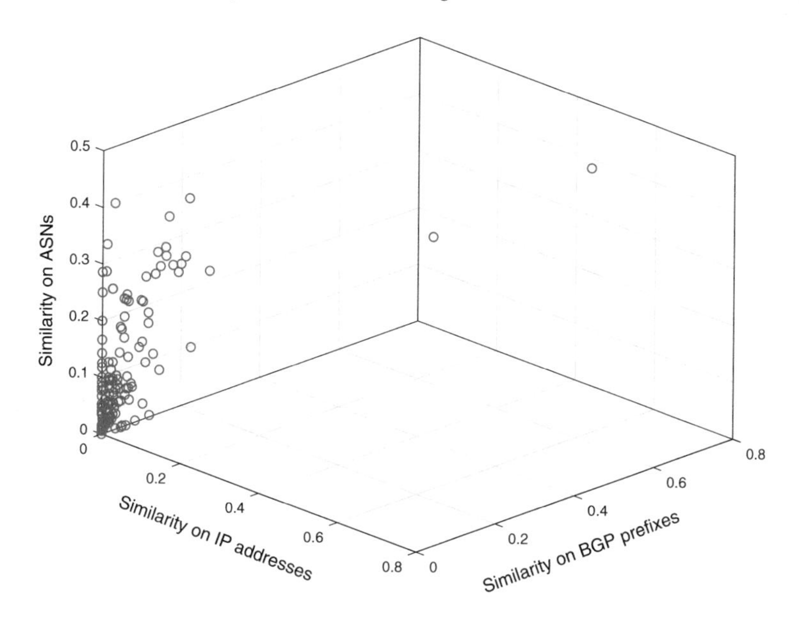

Fig. 8.19 The scatter plot of similarity score on IP-spatial features

networks. As shown in Fig. 8.19, most pairs of IoT devices exhibit low similarities, suggesting IoT devices communicating with diverse servers on the Internet. However, the high similarities between two pairs of IoT devices from two different edge networks are apparently worth in-depth investigations. Our further analysis discovers that two pairs of IoT devices are exactly the same IoT products, i.e., Amazon Echo Dot and Samsung SmartThings Hub, which happen to be deployed in both edge networks. In addition to the similarity scores on IP-spatial features, we also compare the scores on temporal and service dimensions. After ranking the average similarity score over all features, we find that the top pairs of IoT devices with the highest similarity scores, i.e., 0.65 and 0.47, are exactly the same two pairs of devices. We believe that the discovery of high similarity scores on behavioral features among similar IoT devices could help identify newly added or unknown IoT devices by monitoring and learning their behavioral fingerprints during the early phase after they join the edge networks.

Several recent studies have explored machine learning techniques for IoT device detection and classification [75, 77]. For examples, [77] presents a Random Forest classifier to automatically identify device types of the new IoT devices that are connected to a network for the enforcement of security polices and traffic rules, and [75] leverages the widely used supervised classification algorithm, i.e., Random Forest, for classifying authorized and unauthorized IoT devices based on the features extracted from the link and service layers of BLE protocol stacks. The multidimensional behavioral profiles of IoT devices we have developed in this study will provide additional features and unique insights for improving the quality and performance of these machine learning-based IoT device detection and classification.

8.3.5.3 Network Security Monitoring

In light of prevalent cyber attacks and exploits towards vulnerable IoT devices, it is crucial to develop effectively techniques for monitoring traffic activities of IoT devices for network security monitoring. Similar to a network telescope, our proposed measurement framework on programmable edge networks can build the fine-grained and multidimensional behavioral profiles of IoT devices, and provide critical insights for discovering the potential exploits and attacks towards IoT devices in real time.

To demonstrate the feasibility of our proposed IoT measurement framework for network security monitoring, we simulate all the critical steps of Mirai botnet [84, 93] for infiltrating, infecting, and operating weakly protected IP cameras in a controlled edge network environment. For each of the infiltration, infection, and operation steps, we demonstrate that the behavioral fingerprints left by Mirai botnet traffic reveals many unusual traffic patterns or substantial behavioral deviations that could raise anomalous alerts and security alarms.

During the infiltration step, Mirai first employs a port scan strategy for identifying open ports such as 22, 23, and 2323, and if successful, subsequently attempts to launch a dictionary attack to attempt the logins with 62 default credentials. Clearly the scanning activity and brute-force login process trigger substantial behavioral footprint deviations on the IP-spatial and application dimensions, since the IP address of the remote attacker is from a different network prefix and ASN, and the remote ports used in the scanning are very different from the limited set of applications used by IP cameras. The infection stage also leaves unique behavioral fingerprints on IP-spatial, data volumes, and applications, as the loader, which could be different from the initial scanner, has to transfer the malware image to the compromised IP camera.

During the operation stage, the compromised IP camera, as part of Mirai botnet now, exhibits very unusual attacking behaviors since the device starts to (i) perform port scanning activities, (2) communicate with control and command (C2) servers of Mirai botnet, and eventually (3) launch coordinated distributed denial service attacks (DDoS) towards C2-specified targets such as Dyn DNS infrastructure [84]. All of these malicious traffic activities by the IP camera, a new Mirai bot, leaves significant deviations on the behavioral fingerprint on IP camera, thus our proposed multidimensional behavioral profiling framework for IoT devices could effectively detect, mitigate, and stop such malicious activities.

8.4 Summary

In the last few decades, network behavior analysis has provided critical insights in characterizing and modeling traffic behaviors of end systems and Internet applications in traditional networks such as backbone networks and enterprise networks and emerging networks such as data center networks, home networks, and IoT networks. As all the future next-generation networks such as 5G cellular networks and beyond, vehicle networks, and space networks exchange traffic for data communications, network behavior analysis will continue to play an important role in analyzing and modeling massive data traffic for understanding behavioral patterns of networked systems and Internet applications and for detecting and mitigating anomalous and intrusion activities in these future networks.

References

1. K. Xu, F. Wang, L. Gu, Profiling-as-a-service in multi-tenant cloud computing environments, in *Proceedings of International Workshop on Security and Privacy in Cloud Computing* (2012)
2. Amazon: Amazon Web Services, www.aws.amazon.com
3. M. Armbrust, A. Fox, R. Griffith, A.D. Joseph, R. Katz, A. Konwinski, G. Lee, D. Patterson, A. Rabkin, I. Stoica, M. Zaharia, A view of cloud computing. Commun. ACM **53**(4), 50–58 (2010)
4. T. Ristenpart, E. Tromer, H. Shacham, S. Savage, Hey, you, get off of my cloud! exploring information leakage in third-party compute clouds, in *Proceedings of ACM Conference on Computer and Communication Security (CCS)* (2009)
5. D. Owens, Securing elasticity in the cloud. Commun. ACM **53**(6), 46–51 (2010)
6. L. Ertaul, S. Singhal, G. Saldamli, Security challenges in cloud computing, in *Proceedings of International Conference on Security and Management* (2010)
7. Y. Chen, V. Paxson, R. Katz, What<92>s New About Cloud Computing Security? Technical Report No. UCB/EECS-2010-5, University of California at Berkely (2010)
8. M. Yildiz, J. Abawajy, T. Ercan, A. Bernoth, A layered security approach for cloud computing infrastructure, in *Proceedings of International Symposium on Pervasive Systems, Algorithms, and Networks (ISPAN)* (2009)
9. N. Santos, K. Gummadi, R. Rodrigues, Towards trusted cloud computing, in *Proceedings of USENIX Workshop On Hot Topics in Cloud Computing (HotCloud)* (2009)
10. M. Al-Fares, A. Loukissas, A. Vahdat, A scalable, commodity data center network architecture, in *Proceedings of ACM SIGCOMM* (2008)
11. A. Greenberg, J. Hamilton, N. Jain, S. Kandula, C. Kim, P. Lahiri, D.A. Maltz, P. Pat, VL2: a scalable and flexible data center network, in *Proceedings of ACM SIGCOMM* (2009)
12. C. Guo, G. Lu, D. Li, H. Wu, X. Zhang, Y. Shi, C. Tian, Y. Zhang, S. Lu, BCube: a high performance, server-centric network architecture for modular data centers, in *Proceedings of ACM SIGCOMM* (2009)
13. T. Benson, A. Anand, A, Akella, M. Zhang, Understanding data center traffic characteristics, in *Proceedings of SIGCOMM Workshop: Research on Enterprise Networking (WREN)* (2009)
14. H. Jiang, Z. Ge, S. Jin, J. Wang, Network prefix-level traffic profiling: characterizing, modeling, and evaluation. Comput. Netw. (2010)
15. K. Xu, Z.-L. Zhang, S. Bhattacharyya, Internet traffic behavior profiling for network security monitoring. IEEE/ACM Trans. Netw. **16**, 1241–1252 (2008)

16. M. Iliofotou, B. Gallagher, T. Eliassi-Rad, G. Xie, M. Faloutsos, Profiling-by-association: a resilient traffic profiling solution for the internet backbone, in *Proceedings of ACM International Conference on Emerging Networking EXperiments and Technologies (CoNEXT)* (2010)
17. T. Karagiannis, K. Papagiannaki, M. Faloutsos, BLINC: multilevel traffic classification in the dark, in *Proceedings of ACM SIGCOMM* (2005)
18. Y. Hu, D.-M. Chiu, J. Lui, Profiling and identification of P2P traffic. Comput. Netw. **53**(6) (2009)
19. Amazon: Amazon Web Services: Overview of Security Processes, http://aws.amazon.com/security
20. DShield.org: Cooperative Network Security Community - Internet Security, http://www.dshield.org/
21. W. Liu, R. Boutaba, pMeasure: a peer-to-peer measurement infrastructure for the internet. Comput. Commun. **29**(10), 1665–1674 (2006)
22. S. Katti, B. Krishnamurthy, D. Katabi, Collaborating against common enemies, in *Proceedings of ACM SIGCOMM Internet Measurement Conference* (2005)
23. E. Wustrow, M. Karir, M. Bailey, F. Jahanian, G. Houston, Internet background radiation revisited, in *Proceedings of ACM SIGCOMM Conference on Internet Measurement* (2010)
24. B. Bloom, Space/time trade-offs in hash coding with allowable errors. Commun. ACM **13**(7), 422–426 (1970)
25. G. Ren, E. Tune, T. Moseley, Y. Shi, S. Rus, R. Hundt, Google-wide profiling: a continuous profiling infrastructure for data centers. IEEE Micro **30**(4), 65–79 (2010)
26. N. World, DDoS attack against Bitbucket darkens Amazon cloud, http://www.networkworld.com/community/node/45891
27. J. Jung, B. Krishnamurthy, M. Rabinovich, Flash crowds and denial of service attacks: characterization and implications for CDNs and Web sites, in *Proceedings of International World Wide Web Conference* (2002)
28. C. Zou, L. Gao, W. Gong, D. Towsley, Monitoring and early warning for internet worms, in *Proceedings of ACM Conference on Computer and Communications Security* (2003)
29. J. Mai, A. Sridharan, C.-N. Chuah, H. Zang, T. Ye, Impact of packet sampling on portscan detection. IEEE J. Sel. Areas Commun. **24**(12), 2285–2298 (2006)
30. CNN: Netflix takes up 32.7 of Internet bandwidth, http://www.cnn.com/2011/10/27/tech/web/netflix-internet-bandwith-mashable/index.html
31. B. Aggarwal, R. Bhagwan, T. Das, S. Eswaran, V. Padmanabhan, G. Voelker, Unbiased sampling in directed social graph, in *Proceedings of ACM SIGCOMM* (2010)
32. E. Poole, K. Edwards, L. Jarvis, The home network as a sociotechnical system: understanding the challenges of remote home network problem diagnosis. J. Comput. Support. Coop. Work special issue on CSCW, Technology, and Diagnostic Work (2009)
33. J. Yang, W.K. Edwards, D. Haslem, Eden: supporting home network management through interactive visual tools, in *Proceedings of ACM Symposium on User Interface Software and Technology* (2010)
34. R. Grinter, K. Edwards, M. Chetty, E. Poole, J. Sung, J. Yang, A. Crabtree, P. Tolmie, T. Rodden, C. Greenhalgh, S. Benford, The ins and outs of home networking: the case for useful and usable domestic networking. ACM Trans. Comput. Hum. Interact. **16**(2) (2009)
35. R. Grinter, K. Edwards, M. Newman, N. Ducheneaut, The work to make a home network work, in *Proceedings of European Conference on Computer-Supported Cooperative Work (ECSCW)* (2005)
36. G. Maier, A. Feldmann, V. Paxson, M. Allman, On dominant characteristics of residential broadband internet traffic, in *Proceedings of ACM Internet Measurement Conference (IMC)* (2009)
37. K. Cho, K. Fukuda, H. Esaki, A. Kato, The impact and implications of the growth in residential user-to-user traffic, in *Proceedings of ACM SIGCOMM* (2006)
38. M. Dischinger, A. Haeberlen, K. Gummadi, S. Saroiu, Characterizing residential broadband networks, in *Proceedings of ACM Internet Measurement Conference (IMC)* (2007)

39. W. Edwards, R. Grinter, R. Mahajan, D. Wetherall, Advancing the state of home networking. Commun. ACM **54**(6), 62–71 (2011)
40. N. Feamster, Outsourcing home network security, in *Proceedings of ACM SIGCOMM Workshop on Home Networks (HomeNets)* (2010)
41. J. Yang, W.K. Edwards, A study on network management tools of householders, in *Proceedings of ACM SIGCOMM Workshop on Home Networks (HomeNets)* (2010)
42. K. Calvert, W.K. Edwards, N. Feamster, R.E. Grinter, Y. Deng, X. Zhou, Instrumenting home networks, in *Proceedings of ACM SIGCOMM Workshop on Home Networks (HomeNets)* (2010)
43. Y. Yiakoumis, K. Yap, S. Katti, G. Parulkar, N. McKeown, Slicing home networks, in *Proceedings of ACM SIGCOMM Workshop on Home Networking* (2011)
44. L. DiCioccio, R. Teixeira, C. Rosenberg, Impact of home networks on end-to-end performance: controlled experiments, in *Proceedings of ACM SIGCOMM Workshop on Home Networks* (2010)
45. C. Kreibich, N. Weaver, B. Nechaev, V. Paxson, Netalyzr: illuminating the edge network, in *Proceedings of ACM Internet Measurement Conference (IMC)* (2010)
46. M. Chetty, M. Haslem, A. Baird, U. Ofoha, B. Sumner, R. Grinter, Why is my internet slow?: making network speeds visible, in *Proceedings of ACM Conference on Computer-Human Interaction* (2011)
47. S. Hatonen, A. Nyrhinen, L. Eggert, S. Strowes, P. Sarolahti, M. Kojo, An experimental study of home gateway characteristics, in *Proceedings of ACM Internet Measurement Conference* (2010)
48. S. Sundaresan, W. de Donato, N. Feamster, R. Teixeira, S. Crawford, A. Pescape, Broadband internet performance: a view from the gateway, in *Proceedings of ACM SIGCOMM* (2011)
49. K. Xu, F. Wang, L. Gu, J. Gao, Y. Jin, Characterizing home network traffic: an inside view. Accepted by Pers. Ubiquit. Comput. **18**(4), 967–975 (2014)
50. P. De Lutiis, Managing home networks security challenges: security issues and countermeasures, in *Proceedings of International Conference on Intelligence in Next Generation Networks* (2010)
51. OpenWrt: OpenWrt: a Linux distribution for embedded devices, https://openwrt.org/
52. A. Lakhina, K. Papagiannaki, M. Crovella, C. Diot, E. Kolaczyk, N. Taft, Structural analysis of network traffic flows, in *Proceedings of ACM SIGMETRICS* (2004)
53. A. Lakhina, M. Crovella, C. Diot, Diagnosing network-wide traffic anomalies, in *Proceedings of ACM SIGCOMM* (2004)
54. K. Xu, J. Chandrashekar, Z.-L. Zhang, Principal component analysis on BGP update streams. J. Commun. Netw. **12**(2), 191–197 (2010)
55. K. Xu, J. Chandrashekar, Z.L. Zhang, A first step towards understanding inter-domain routing, in *Proceedings of ACM SIGCOMM Workshop on Mining Network Data* (2005)
56. I.T. Jolliffe, *Principal Component Analysis*, 2nd edn. Springer Series in Statistics (2002)
57. R. Pang, V. Yegneswaran, P. Barford, V. Paxson, L. Peterson, Characteristics of internet background radiation, in *Proceedings of ACM Internet Measurement Conference* (2004)
58. R. Want, B. Schilit, S. Jenson, Enabling the internet of things. Computer **48**(1), 28–35 (2015)
59. S. Shams, S. Goswami, K. Lee, S. Yang, S.-J. Park, Towards distributed cyberinfrastructure for smart cities using big data and deep learning technologies, in *Proceedings of IEEE International Conference on Distributed Computing Systems (ICDCS)* (2018)
60. W. Shi, J. Cao, Q. Zhang, Y. Li, L. Xu, Edge computing: vision and challenges. IEEE Internet Things J. **3**(5), 637–646 (2016)
61. E. Fernandes, J. Paupore, A. Rahmati, D. Simionato, M. Conti, A. Prakash, FlowFence: practical data protection for emerging IoT application frameworks, in *Proceedings of USENIX Conference on Security Symposium* (2016)
62. E. Ronen, A. Shamir, A.-O. Weingarten, C. O'Flynn, IoT goes nuclear: creating a ZigBee chain reaction, in *Proceedings of IEEE Symposium on Security and Privacy (S&P)* (2017)
63. K. Xu, F. Wang, S. Jimenez, A. Lamontagne, J. Cummings, M. Hoikka, Characterizing DNS behaviors of internet-of-things in edge networks. IEEE Internet Things J. **7**(9) (2020)

64. Q. Wang, W. Hassan, A. Bates, C. Gunter, Fear and logging in the internet of things, in *Proceedings of Network and Distributed System Security Symposium (NDSS)* (2018)
65. Y. Wan, K. Xu, F. Wang, G. Xue, IoTArgos: a multi-layer security monitoring system for internet-of-things in smart homes, in *Proceedings of IEEE INFOCOM* (2020)
66. A. Zanella, N. Bui, A. Castellani, L. Vangelista, M. Zorzi, Internet of things for smart cities. IEEE Internet Things J. **1**(1), 22–32 (2014)
67. E. Fernandes, J. Jung, A. Prakash, Security analysis of emerging smart home applications, in *Proceedings of IEEE Symposium on Security and Privacy (S&P)* (2016)
68. G. Ho, D. Leung, P. Mishra, A. Hosseini, D. Song, D. Wagner, Smart locks: lessons for securing commodity internet of things devices, in *Proceedings of ACM on Asia Conference on Computer and Communications Security (ASIACCS)* (2016)
69. K. Xu, F. Wang, X. Jia, Secure the internet, one home at a time. Secur. Commun. Netw. **9**(16), 3821–3832 (2016)
70. K. Xu, Y. Wan, G. Xue, Powering smart homes with information-centric networking. IEEE Commun. Mag. **57**(6) (2019)
71. S. Feng, P. Setoodeh, S. Haykin, Smart home: cognitive interactive people-centric internet of things. IEEE Commun. Mag. **55**(2), 34–39 (2017)
72. W. Zhang, Y. Meng, Y. Liu, X. Zhang, Y. Zhang, H. Zhu, HoMonit: monitoring smart home apps from encrypted traffic, in *Proceedings of ACM SIGSAC Conference on Computer and Communications Security (CCS)* (2018)
73. K. Gao, C. Corbett, R. Beyah, A passive approach to wireless device fingerprinting, in *Proceedings of IEEE/IFIP International Conference on Dependable Systems & Networks (DSN)* (2010)
74. S. Siby, R. Maiti, N. Tippenhauer, IoTScanner: detecting privacy threats in IoT neighborhoods, in *Proceedings of ACM International Workshop on IoT Privacy, Trust, and Security (IoTPTS)* (2017)
75. T. Gu, P. Mohapatra, BF-IoT: securing the IoT networks via fingerprinting-based device authentication, in *Proceedings of IEEE International Conference on Mobile Ad Hoc and Sensor Systems (MASS)* (2018)
76. B. Bezawada, M. Bachani, J. Peterson, H. Shirazi, I. Ray, I. Ray, Behavioral fingerprinting of IoT devices, in *Proceedings of ACM CCS Workshop on Attacks and Solutions in Hardware Security (ASHES)* (2018)
77. M. Miettinen, S. Marchal, I. Hafeez, N. Asokan, A.-R. Sadeghi, S. Tarkoma, IoT SENTINEL: automated device-type identification for security enforcement in IoT, in *Proceedings of IEEE International Conference on Distributed Computing Systems (ICDCS)* (2017)
78. H. Jafari, O. Omotere, D. Adesina, H.-H. Wu, L. Qian, IoT devices fingerprinting using deep learning, in *Proceedings of IEEE Military Communications Conference (MILCOM)* (2018)
79. K. Xu, Y. Wan, G. Xue, F. Wang, Multidimensional behavioral profiling of internet-of-things in edge networks, in *Proceedings of IEEE/ACM International Symposium on Quality of Service (IWQoS)* (2019)
80. Y. Wan, K. Xu, F. Wang, G. Xue, Characterizing and mining traffic patterns of IoT devices in edge networks. IEEE Trans. Netw. Sci. Eng. **8**(1) (2021)
81. Business Insider: Google, This is Bogus as Hell - One of the Fathers of the Internet Blasts Google for how Chromecast Behaves on His Home Network (2019), https://www.businessinsider.com/paul-vixie-blasts-google-chromecast-2019-2/
82. Y. Rekhter, T. Li, A Border Gateway Protocol 4 (BGP-4) (1995)
83. O. Alrawi, C. Lever, M. Antonakakis, F. Monrose, SoK: security evaluation of home-based IoT deployments, in *Proceedings of IEEE Symposium on Security and Privacy (S&P)* (2019)
84. M. Antonakakis, T. April, M. Bailey, M. Bernhard, E. Bursztein, J. Cochran, Z. Durumeric, J. Halderman, L. Invernizzi, M. Kallitsis, D. Kumar, C. Lever, Z. Ma, J. Mason, D. Menscher, C. Seaman, N. Sullivan, K. Thomas, Y. Zhou, Understanding the Mirai Botnet, in *Proceedings of USENIX Security Symposium* (2017)
85. J. Zhang, X. Chen, Y. Xiang, W. Zhou, J. Wu, Robust network traffic classification. IEEE/ACM Trans. Netw. **23**(4), 1257–1270 (2015)

86. M. Trevisan, D. Giordano, I. Drago, M. Mellia, M. Munafo, Five years at the edge: watching internet from the ISP network, in *Proceedings of ACM International Conference on Emerging Networking EXperiments and Technologies (CoNEXT)* (2018)
87. of Oregon, U.: Routeviews archive project, http://archive.routeviews.org/
88. K. Park, W. Willinger, *Self-Similar Network Traffic and Performance Evaluation* (Wiley, 2002)
89. Y. Meidan, M. Bohadana, A. Shabtai, J. Guarnizo, M. Ochoa, N. Tippenhauer, Y. Elovici, ProfilIoT: a machine learning approach for IoT device identification based on network traffic analysis, in *Proceedings of ACM Symposium on Applied Computing* (2017)
90. L. Akoglu, H. Tong, J. Vreeken, C. Faloutsos, Fast and reliable anomaly detection in categorical data, in *Proceedings of ACM international conference on Information and knowledge management (CIKM)* (2012)
91. M. Li, P. Vitányi, *An Introduction to Kolmogorov Complexity and its Applications* (Springer, 1993)
92. P. Grünwald, *The Minimum Description Length Principle* (MIT Press, 2007)
93. G. Kambourakis, C. Kolias, A. Stavrou, The Mirai botnet and the IoT Zombie Armies, in *Proceedings of IEEE Military Communications Conference (MILCOM)* (2017)

Printed in the United States
by Baker & Taylor Publisher Services